Walk Around

A-7 Corsair II

By Lou Drendel

Color by Lou Drendel and Don Greer

Squadron/Signal® Publications

Walk Around Number 44

Introduction

The Ling-Temco-Vought (LTV) A-7 Corsair II was one of the most successful military aircraft of modern times and probably the greatest bargain in the history of weapon systems procurement. A-7As cost a little over one million dollars each and delivered weapons with an accuracy unheard of in their day while achieving the lowest loss rate of any aircraft in the Vietnam war. Later models went on to pioneer the sophisticated avionics systems which are the norm today. The head-up display, the central navigation-weapon delivery computer, and other avionics innovations were first used on the A-7.

The life of the A-7 covers several periods. The A-7A, A-7B, and A-7C were basically the same airplane with engine updates and other engine-related system modifications. The A-7D and A-7E were giant steps into the future, with sophisticated avionics suites which set the pattern for all future weapon delivery and navigation systems. The next phase was a series of two-place designs, encompassing the US Navy's TA-7C, the US Air Force's A-7K, Greece's TA-7H, and Portugal's TA-7P. These were followed by the low-altitude night attack (LANA) version of the A-7D and A-7K. A major engine upgrade took the A-7 into the supersonic performance range as the US Air Force's YA-7F.

The A-7A first flew in September 1965, and the first operational US Navy squadron airplane was delivered in October 1966. A-7As entered combat with the Navy in December 1967, and the Navy flew its final A-7E combat missions during the Gulf War in 1991. The A-7Ds and A-7Ks of the US Air Force were retired from active service in 1993, with final operations still being conducted by Air National Guard units.

LTV manufactured 1,545 A-7 airframes. Of these, 113 were remanufactured to produce additional models of the airplane. These included sixty TA-7Cs, forty-four A-7Bs, six TA-7Ps, one YA-7H/E, and two YA-7Fs. Between 1968 and 1991, A-7s logged over five million flight hours and were the US military's most cost-effective aerial weapon. The Corsair II was noted for its ability to carry very heavy weapon loads (up to 20,000 pounds), low maintenance requirements (nine to eleven maintenance man-hours per flight hour), superior weapon delivery accuracy (bombing within 50 meters of friendly troops), long range (up to 4,250 nautical miles), a low loss rate in combat (0.04 percent), and a very low accident rate. A-7s are still in service in Portugal, Greece, and Thailand.

Acknowledgements

Doing a *Walk Around* on an airplane which is no longer operational is never easy. Many older airplanes exist within the warbird community, but recent jets are particularly difficult to access if they don't exist in a museum. Fortunately, the National Museum of Naval Aviation, on NAS Pensacola, Florida, has A-7s on display, and they allowed me unfettered access to photograph the details.

A number of photographers were generous in allowing me to use their work. I am particularly indebted to Ted Carlson, Andre Jans, Peter Steendam, Jorge Manuel Antao Ruivo, Dave Mason, Don Logan, Dr. J.G. Handelman, Bruce Trombecky, Colin Norwood, Ken Buchanan, and Den Pascoe.

ISBN 0-89747-508-9

If you have any photographs of aircraft, armor, soldiers or ships of any nation, particularly wartime snapshots, why not share them with us and help make Squadron/Signal's books all the more interesting and complete in the future? Any photograph sent to us will be copied and the original returned. The donor will be fully credited for any photos used. Please send them to:

Squadron/Signal Publications
1115 Crowley Drive
Carrollton, TX 75006

Если у вас есть фотографии самолётов, вооружения, солдат или кораблей любой страны, особенно, снимки времён войны, поделитесь с нами и помогите сделать новые книги издательства Эскадрон/Сигнал ещё интереснее. Мы переснимем ваши фотографии и вернём оригиналы. Имена приславших снимки будут сопровождать все опубликованные фотографии. Пожалуйста, присылайте фотографии по адресу:

Squadron/Signal Publications
1115 Crowley Drive
Carrollton, TX 75006

軍用機、装甲車両、兵士、軍艦などの写真を所持しておられる方はいらっしゃいませんか？どの国のものでも結構です。作戦中に撮影されたものが特に良いのです。Squadron/Signal社の出版する刊行物において、このような写真は内容を一層充実し、興味深くすることができます。当方にお送り頂いた写真は、複写の後お返しいたします。出版物中に写真を使用した場合は、必ず提供者のお名前を明記させて頂きます。お写真は下記にご送付ください。

Squadron/Signal Publications
1115 Crowley Drive
Carrollton, TX 75006

Front Cover: An A-7E of VA-113 'Stingers' while assigned to USS *Ranger* (CV 61).

Title page: A pair of A-7Es of VA-72 trail vapor from wingtips. Low altitude, high humidity, and elevated angle of attack create this phenomenon.

Back Cover: An A-7D of the 354th TFW during Linebacker and Linebacker II combat operations in Vietnam, 1972. The 354th commander's airplane was the only A-7 of that unit to carry the sharkmouth.

▲ An A-7A of reserve squadron VA-203. A total of 193 A-7As were built. The A-7A was powered by the Pratt & Whitney TF-30-P-6 non-afterburning turbofan engine, which produced 11,350 pounds of static thrust.

▼ The A-7A was armed with a pair of Mk 12 Mod 0 20 mm guns, which were pneumatically charged, fed, and safed, but used electrical power for firing and control of the pneumatic functions. Expended shell casings were expelled into a compartment below each gun.

▼ Gun ports were located on either side of the engine intake. Gun bays and expended shell casing compartments were ventilated with engine bleed air to remove explosive gases. The guns could not be fired with the landing gear handle in the down position. (Ken Buchanan)

▲ The basic size and shape of the A-7's nose did not change from A-7A to A-7K, although notable changes in upgraded A-7Ds included the addition of a 'Pave Penny' passive laser tracker pod under the intake. (Den Pascoe)

▶ The A-7E was the definitive Navy version of the Corsair II. A principal difference between USAF and Navy versions was their method of air-to-air refueling. Navy models were equipped with a retractable probe, while USAF models were fitted with a receptacle for a 'flying boom.' (US Navy)

▼ Two-seat A-7K (front) and single-seat A-7D did not differ much in frontal area. Aerial refueling receptacles for both were on top of the fuselage.

▲ All naval versions of the Corsair II (A, B, C, and E) were equipped with an air-to-air refueling probe on the starboard side of the nose. This brand-new A-7E was photographed on the LTV ramp in 1975.

◄ All versions of the Corsair II contain an integral boarding ladder and fold-down steps. Red warning triangles caution crew members about the pyrotechnic ejection seat.

▼ The nose radome of the A-7A housed the AN/APQ-116 radar antenna. The Pitot tube is on the port side of the nose just aft of the radome. Vents on the port side of the nose of all versions provide cooling for electronic equipment. Just below the aft canopy corner is the angle-of-attack indexer vane.

A-7E

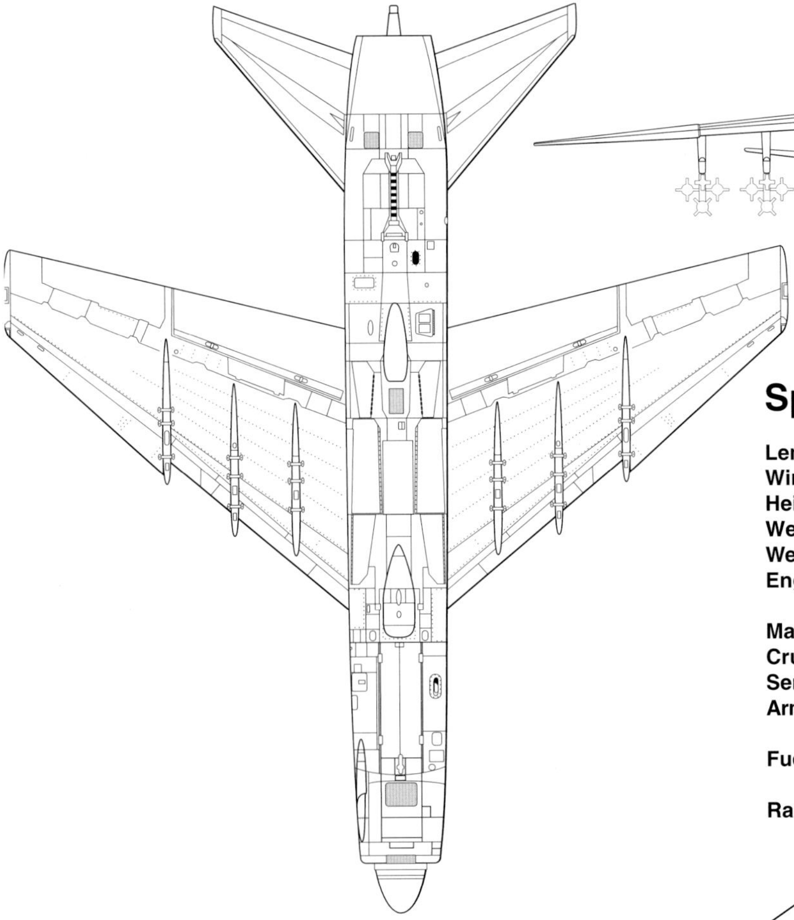

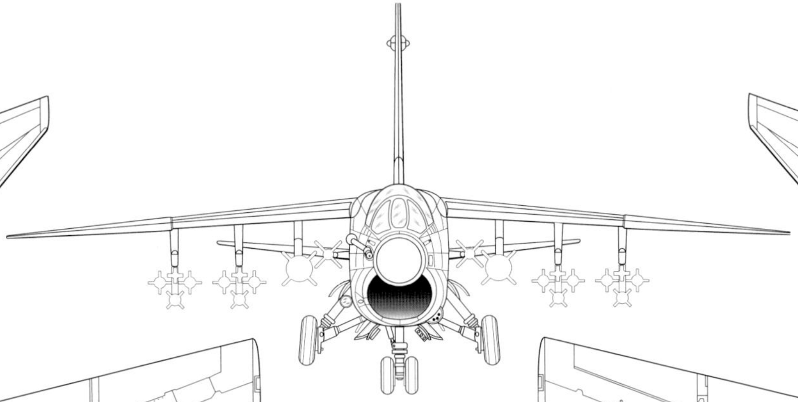

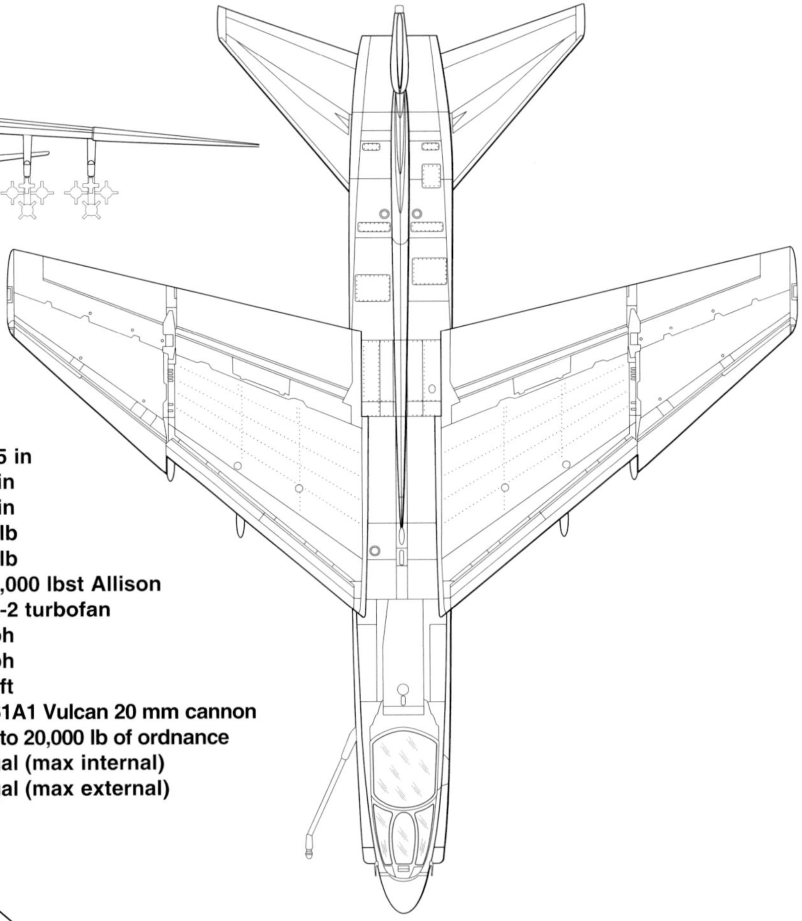

Specifications

Length	46 ft 1.5 in
Wingspan	38 ft 9 in
Height	16 ft 2 in
Weight, empty	19,127 lb
Weight, max loaded	42,000 lb
Engine	One 15,000 lbst Allison TF41-A-2 turbofan
Maximum speed	693 mph
Cruising speed	535 mph
Service ceiling	47,000 ft
Armament	One M61A1 Vulcan 20 mm cannon and up to 20,000 lb of ordnance
Fuel	1,496 gal (max internal) 1,200 gal (max external)
Range, combat	715 mi

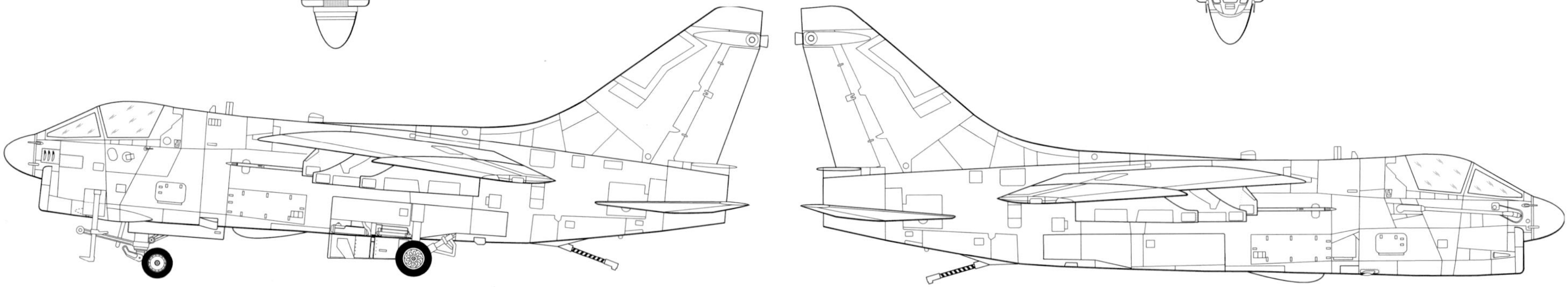

A-7D

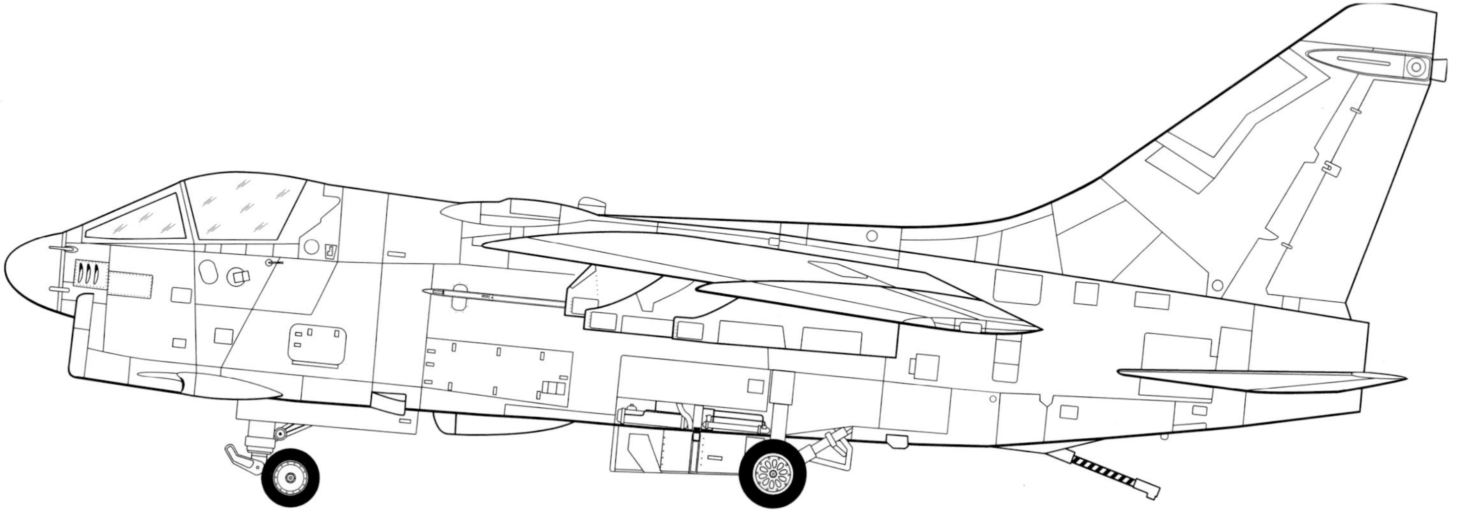

TA-7C

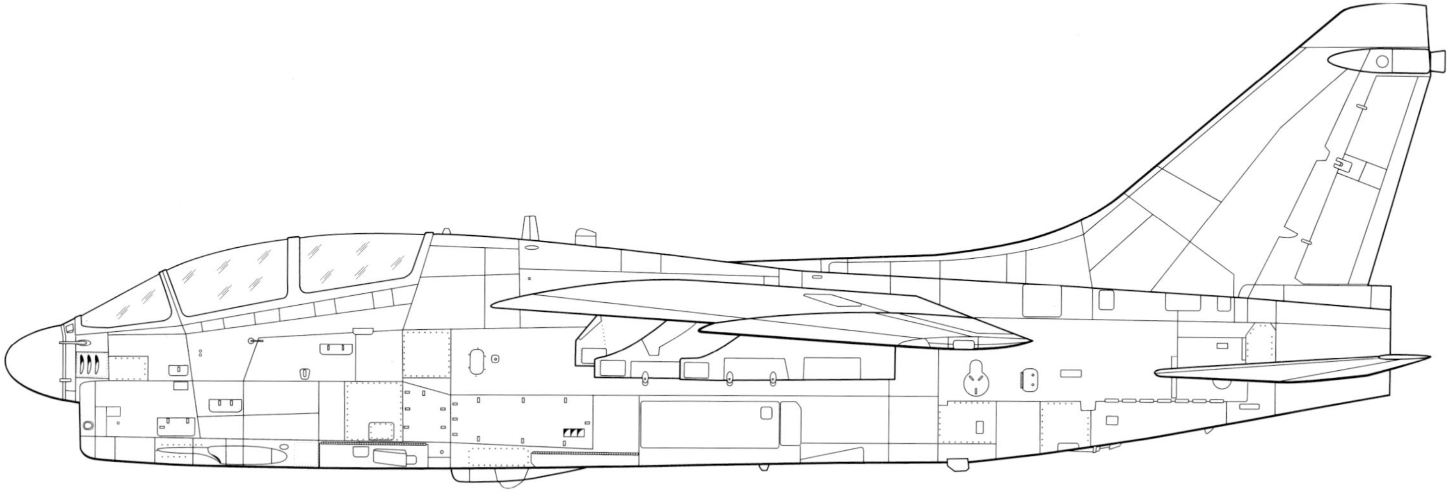

▲ LTV's efforts to sell the Navy a two-seat version of the Corsair II led to the TA-7C. The first production A-7E (BuNo 156801) was bailed to LTV by the Navy, and in a whirlwind design and manufacturing effort, was reborn as the two-seat V-519, which first flew in August 1972. Later it was redesignated YA-7H and ultimately YA-7E. Sixty two-seat production versions, designated TA-7C, were built for the US Navy. (Den Pascoe)

◄ The A-7E's boarding ladder is manually extended. Users are instructed to board the ladder right foot first.

▼ LTV received a contract in 1979 to convert an A-7D to two-seat configuration. Lengthening plugs were inserted in both front and rear fuselage. The new two-seater, designated TA-7D, had full operational capability and could carry the same weapons as the A-7D. The TA-7D was later redesignated A-7K. In addition to the prototype, LTV built thirty new A-7Ks. Unlike the Navy's TA-7Cs, the USAF's A-7Ks were new-builds and were not conversions of earlier single-seaters. Delivery to ANG units began in 1981, and production of the A-7K ended in September of 1984. (Den Pascoe)

An A-7B of VA-174 'Hell Razors' at NAS Cecil Field, Florida. The A-7B differed from the A-7A primarily in having the 12,200 lbst TF30-P-8 engine in place of the -6 version in the A as well as improved flaps. The first A-7B (BuNo 154363) made its first flight on 6 February 1968. A total of 196 A-7Bs were built.

An A-7D seen at Trenton, Ontario, in 1982. Bulges on the upper edges of the intake lip contain AN/APR-36 electronic countermeasure (ECM) warning antennas. The lower (faired) lip of the intake contains an AN/ARN-58A ILS glideslope antenna. (Den Pascoe)

All two-seat A-7s had a canopy hinged on the starboard side. Late model (D and E) A-7s had Pitot tubes on both sides of the nose. This A-7K of the Virginia ANG has an intake cover installed for the static display at an air show. (Den Pascoe)

A-7A General Arrangement

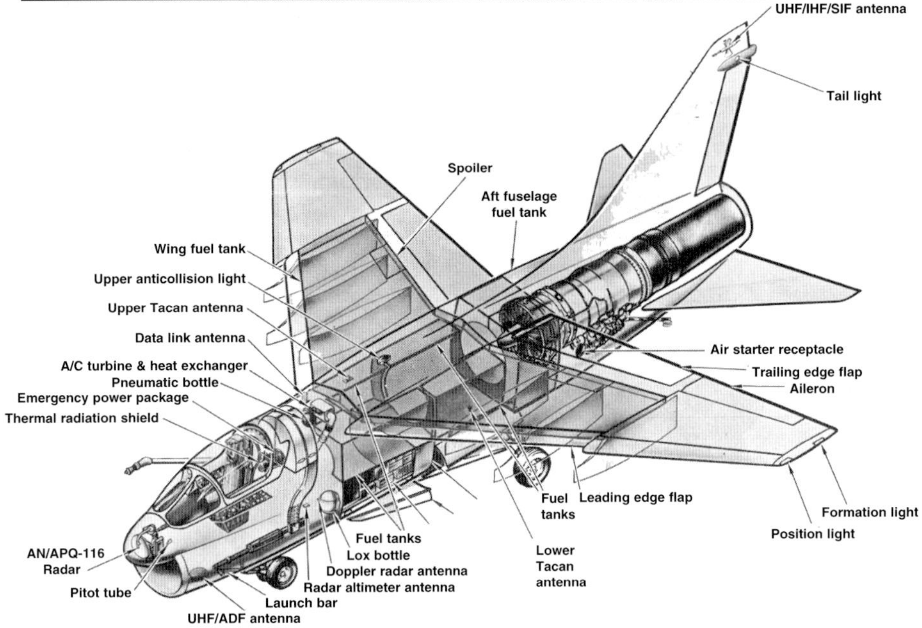

- Major Charles McClarren, USAF, commanded a detachment from the 57th Fighter Weapons Wing, which evaluated the A-7A in combat aboard USS *Ranger* with VA-147 in Project Coronet Stallion. (USN)

A-7D General Arrangement

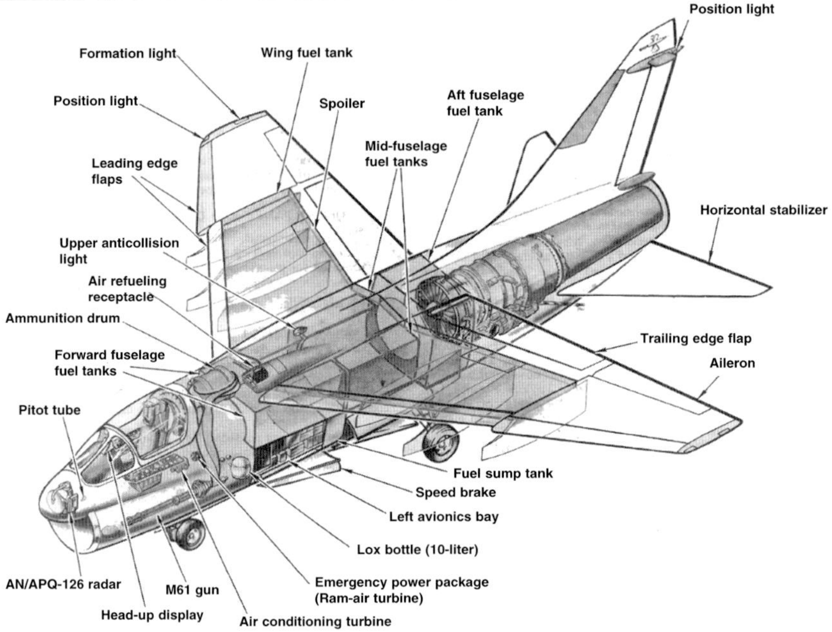

A-7E General Arrangement

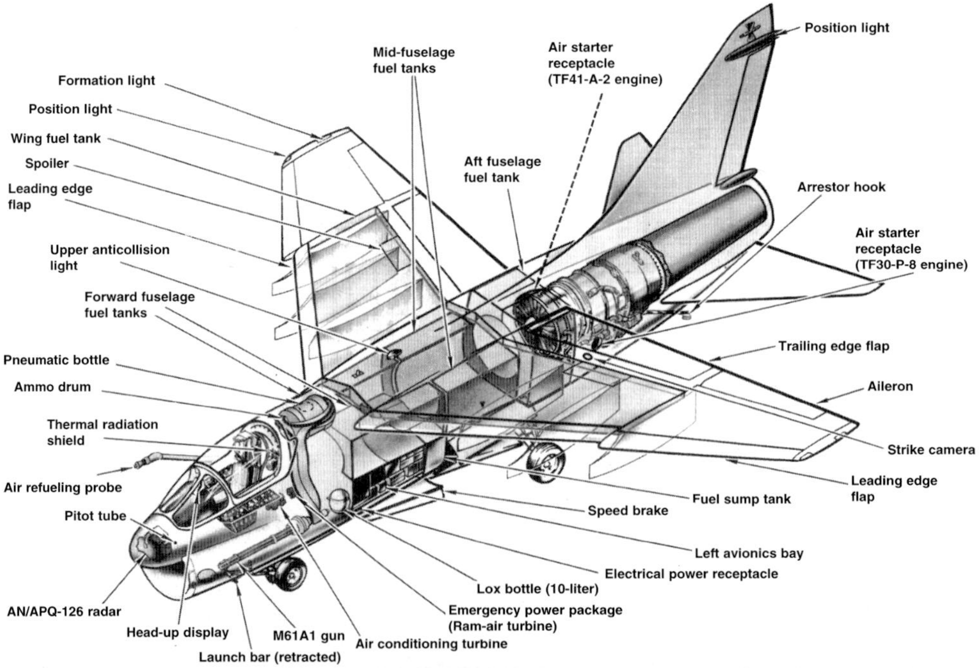

▲ Starboard side of an A-7E forward fuselage on the assembly line. Engine air intake duct framing is visible in open area at left.

▲ Brand-new A-7E at the LTV factory in Dallas, 1975. One of the notable exterior differences between A, B, and C models and the D, E, and K models is the replacement of the starboard 20 mm cannon with an air intake for cooling electronics in the later models. (Lou Drendel)

▼ Port side of an A-7E forward fuselage on the assembly line. The open panel at lower right shows the liquid oxygen container compartment. (Lou Drendel)

TF-30 Engine

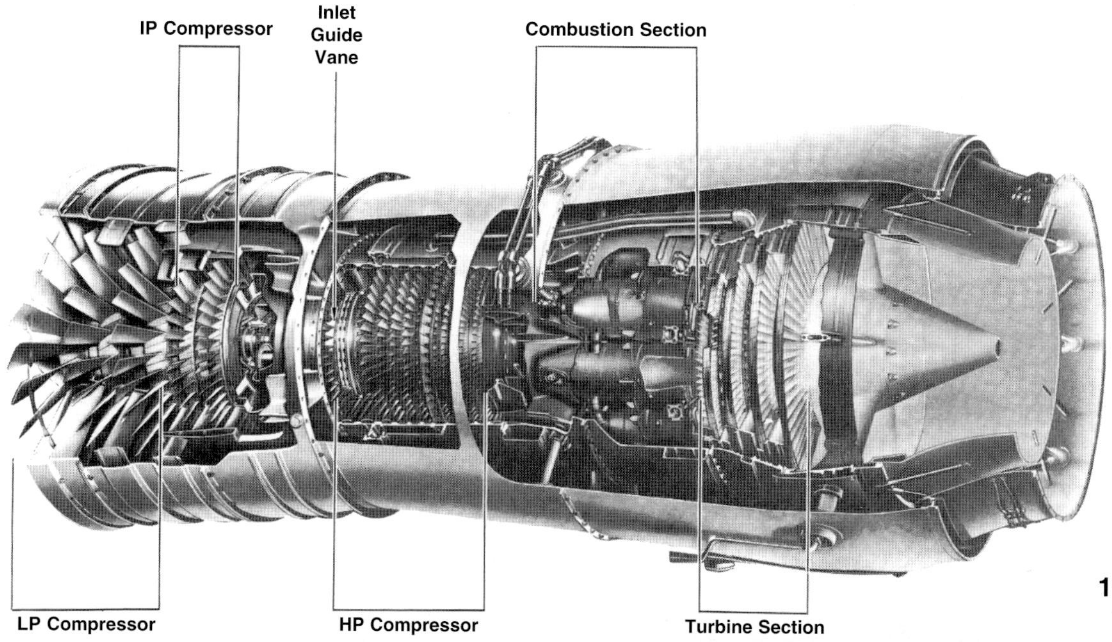

Forward fuselage of TA-7P 15550. The open compartment houses the built-in electronics test panel which allows maintenance personnel to diagnose problems quickly. The Força Aerea Portuguesa (FAP — Portuguese Air Force) has operated a version of the Corsair II, designated A-7P ('P' for Portugal), since the early 1980s. Six TA-7Ps were delivered in May 1985. (Peter Steendam)

▲ Forward end of the refueling probe fairing on the A-7E, with the probe extended and out of sight. Also visible is the red-tipped Pitot tube. (Dave Mason)

▼ The exhaust for the A-7E's air-conditioning turbine is on the lower starboard forward fuselage. (Dave Mason)

▲ Boarding steps in use on A-7D 70-1052 of the Ohio ANG at Trenton, Ontario, Canada in 1983. A slightly worn Distinguished Unit Citation ribbon is carried on the nose just below the windshield. The gun gas vent door on the lower fuselage is open. (Den Pascoe)

▲ The rear of the extended refueling probe, showing the actuating rod. (Dave Mason)

▼ Corsair II fuselage center sections under assembly in the LTV plant. (Lou Drendel)

Antenna Locations

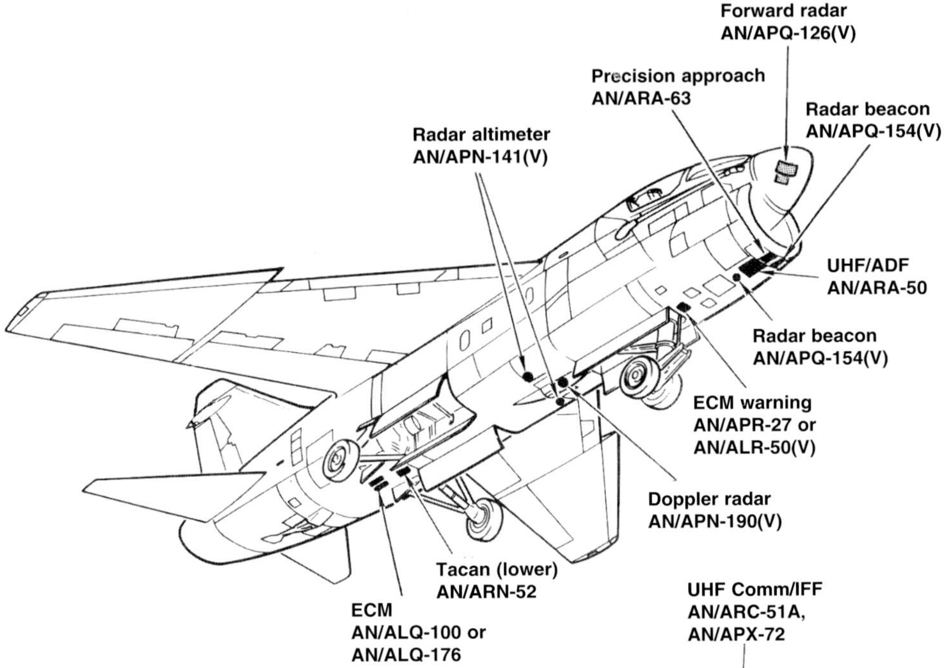

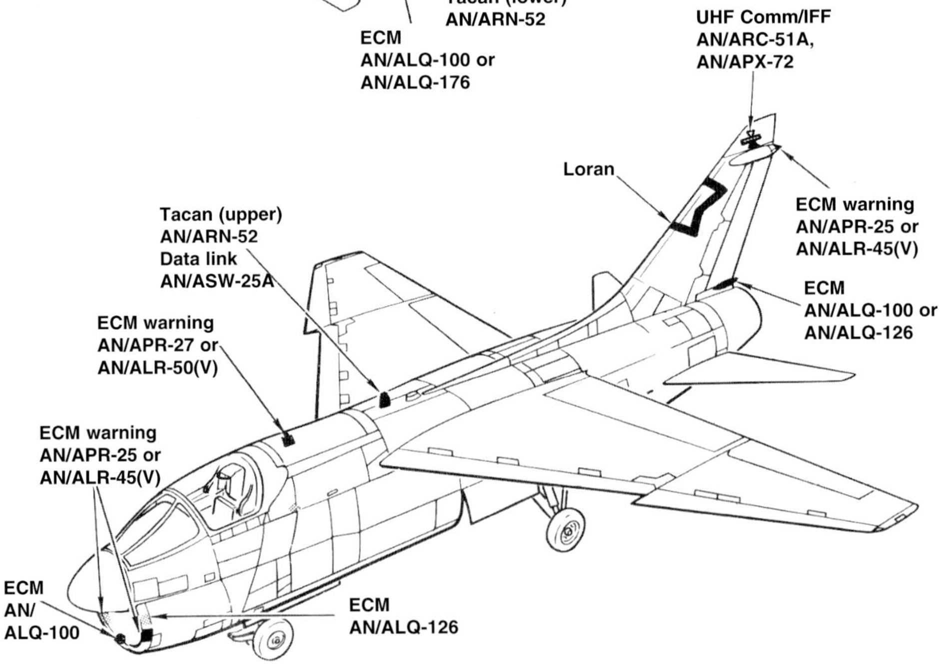

▲ The gun bay venting system uses low-pressure bleed air. When the gun firing trigger switch is squeezed to the second detent, an electrically operated gas purge valve opens, admitting low-pressure air to the gun bay. The ventilating air and explosive gases are vented through this door. (Dave Mason)

▼ Refueling probe extended. The sump tank is filled by gravity flow from the forward and mid-fuselage tanks and by gravity or ejector pump flow from the aft fuselage tank. When the fuselage tanks and the sump tank are full, fuel flows from the forward and aft fuselage tanks into the wing tank. (Dave Mason)

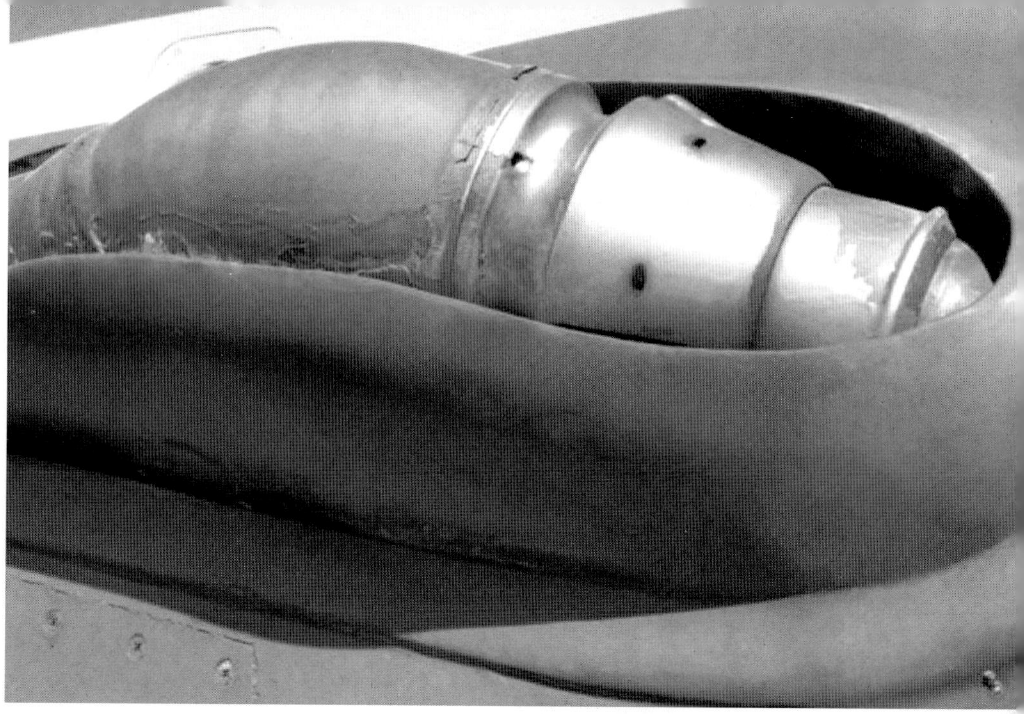

▲ Refueling probe tip in stored position. When refueling is complete, the probe is retracted by hydraulic pressure applied to the probe actuating cylinder.

▼ Vents in both sides of the nose aid in cooling the radar compartment. (Dave Mason)

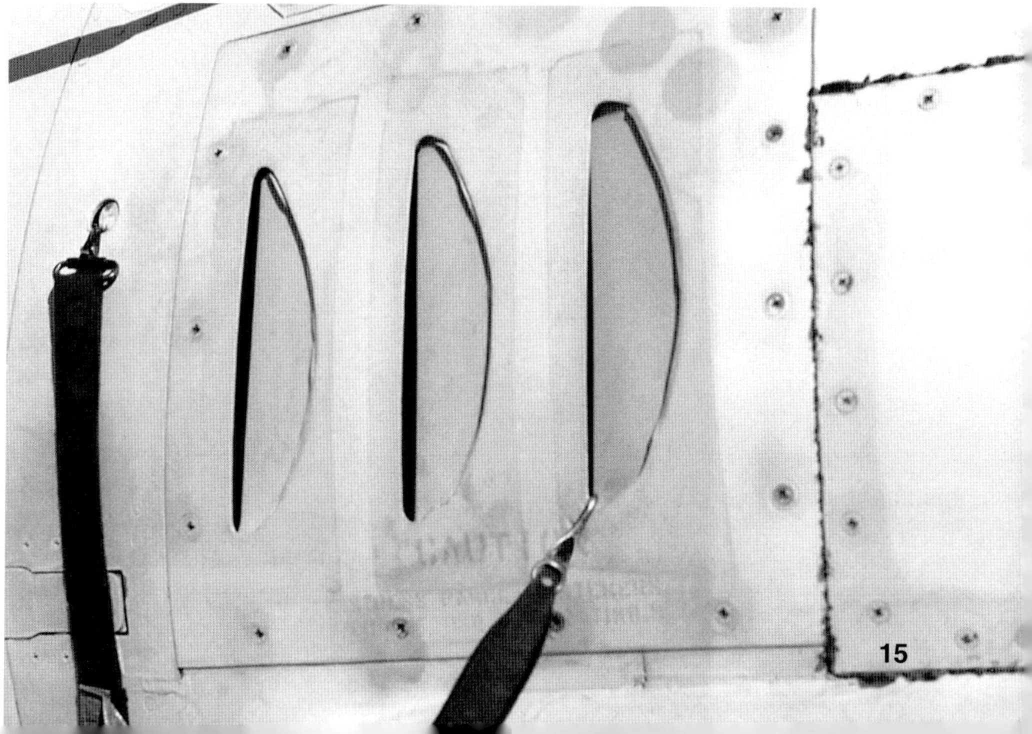

▲ In May 1980, LTV received a contract to convert twenty A-7As into A-7Ps for the Portuguese Air Force. The A-7P was powered by the TF30-P408 engine and was equipped with A-7E avionics. Deliveries began in December 1981. Aircraft number 15521 received this special paint scheme to celebrate sixty-four thousand A-7 hours and was on static display for the 52nd Anniversary of the FAP at Monte Real, Portugal, 10 July 2004. (Jorge Manuel Antao Ruivo)

▶ Though equipped with newer avionics, the A-7P retained the 20 mm cannon of the A-7A. Gun gases were vented from louvers in fuselage, just below the black-outlined armament status rectangle. (Jorge Manuel Antao Ruivo)

▼ A second batch of A-7Ps was ordered in September 1982. Twenty-four A-7Ps and six TA-7Ps were provided. Deliveries of the A-7Ps began in October 1984, when they were issued to the newly-established Esquadrão 303 (303 Squadron). The TA-7Ps followed in May 1985. Prior to that, only a single leased TA-7C was available. (Jorge Manuel Antao Ruivo)

▲ An A-7B of VA-153. Internal armament consisted of two 20 mm Mk 12 cannons with 600 rounds per gun. Up to 15,000 pounds of ordnance could be carried on eight hardpoints. (Charles Howes)

▼ An A-7D at Davis Monthan AFB, Arizona. The A-7D differed from the Navy's Corsair II in several ways. The most significant was the Allison TF41-A-1 turbofan engine, a license-built version of the Rolls-Royce Spey with a thrust of 14,500 pounds, more than 2,000 pounds greater than that of the TF30 that powered the Navy's Corsair IIs. Other changes included a new avionics package, a computerized navigation/weapons delivery system with AN/APQ-126 radar, and a head-up display. (Don Logan)

▼ The impressive performance of the Spey-powered A-7D inspired the Navy to order a new version of their A-7. The A-7E incorporated the Vulcan cannon and the Spey engine used in the A-7D as well as an upgraded avionics suite. The A-7E's first combat cruise was with VA-146 and VA-147, aboard USS *America* in 1970.

Starboard rear fuselage of an A-7E. The large keyhole-shaped access panel in front of the stabilator covers the air start receptacle. Other access panels conceal PC 2 and PC 3 ground test disconnects. The access panel immediately below and behind the wing trailing edge is for inspection of the constant speed drive's oil level. (Lou Drendel)

Port forward fuselage of an A-7E. The large panel under the national insignia provides oxygen servicing access. (Lou Drendel)

Starboard forward fuselage of an A-7E. The primary component behind the access door is the air conditioning turbine unit. (Lou Drendel)

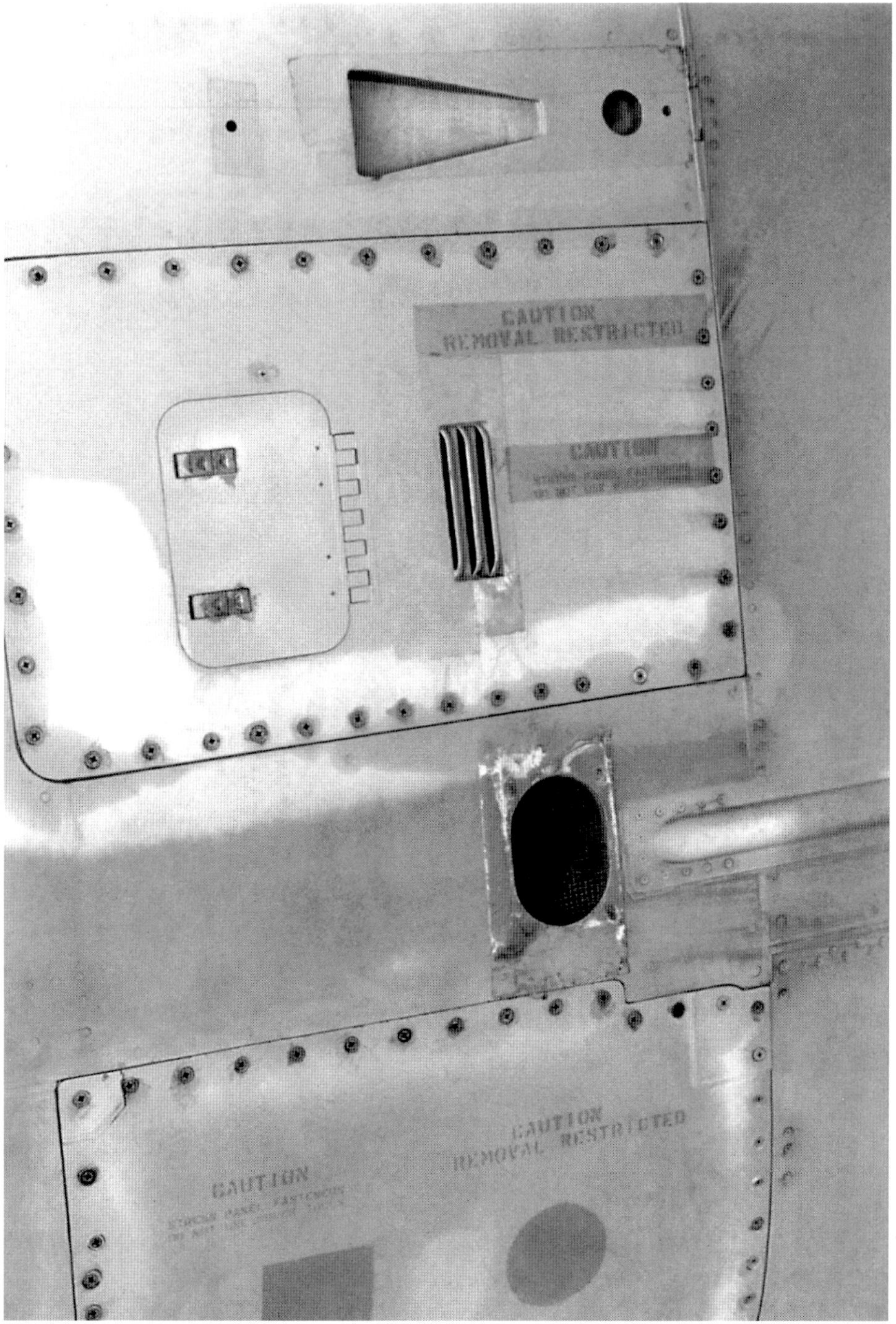

▲ The air conditioning turbine access panel has several cooling vents for this compartment. (Lou Drendel)

◀ Starboard mid-fuselage of an A-7E. The small panel aft of the louvered vent is the constant speed drive oil level inspection access. (Lou Drendel)

▼ The prototype TA-7C, affectionately dubbed 'The White Whale' by LTV employees, shows off the belly-mounted speed brake common to all versions of the A-7. (Vought)

▲ The speed brake is electrically controlled and hydraulically actuated using PC 2 hydraulic pressure. The brake can be extended and held in any position between fully closed and fully open (sixty degrees).

▲ The brake automatically retracts when the landing gear handle is placed in the 'wheels down' position or when electrical power is lost. Airloads will partially retract the brake at high airspeeds.

▼ Vought's first international sale of the A-7 was sixty aircraft to the Hellenic Air Force. Designated A-7H, its most significant difference from the A-7E was the deletion of in-flight refueling and addition of the electrically started gas turbine engine starter employed on the A-7D. Following a first flight on 6 May 1975, A-7H deliveries took place over a two-year period beginning in the summer of 1975. Impressed by the features of the two-place TA-7C, the Greek government placed an order for five of this model, designating it TA-7H. Deliveries took place between July and December 1980. A total of sixty-five A-7s were purchased, and the A-7H fleet is still flying today. (Andre Jans)

▲ Department of Defense strategy called for the ultimate transfer of all A-7Ds to the Air National Guard, which determined that a two-seat combat trainer version of the A-7 was needed to assure the maximum in combat training effectiveness. Designated A-7K, the new trainer retained all of the features of the A-7D and was fully combat-capable. Unlike the Navy's two-seaters, which were converted from single-seat aircraft, the USAF's A-7Ks were all new-builds. In addition to the prototype, Vought built thirty new A-7Ks. Delivery to ANG units began in 1981, and production of the A-7K ended in September of 1984.

▼ TA-7Cs and EA-7Ls of the Naval Air Warfare Center, NAS Point Mugu, California, 1994. The EA-7L was a modified TA-7C with the ability to carry jamming pods and missile simulators on underwing pylons to duplicate Soviet weapons and tactics during training. (Vance Vasquez/NAWCWpns)

▼ EA-7Ls of the Naval Air Weapons Center take off for the last time from NAS Point Mugu on 17 November 1994. These were the Navy's last operational Corsair IIs, and their retirement marked the end of twenty-eight years of the Navy's Corsair II flight operations. (Vance Vasquez/NAWCWpns)

▲ An EA-7L of VAQ-34. Upon establishment of this fleet adversary squadron, six TA-7Cs were transferred to the squadron from existing fleet assets. The planes were modified at NAS Point Mugu, where change kits were installed to convert the aircraft to electronic warfare platforms to create a realistic air-launched threat scenario. They were redesigned EA-7Ls to indicate their unique configuration. (Ted Carlson)

▼ An A-7E of VA-174. During the 1980s, Navy tactical aircraft markings reverted to the muted tones of wartime, as evidenced by this 1986 display at London, Ontario. (Den Pascoe)

▼ An A-7B of VA-153 while assigned to USS *Oriskany*. The last *Oriskany* mission was flown 27 January 1973 by Commander Denis R. Weichman of VA-153. It was his 612th mission of the war — a number greater than that of any other US Naval aviator. (Bruce Trombecky)

▲ An A-7E of VA-27. One of the notable differences between B and E models is the addition of an ECM antenna at the base of the rudder of the E model. (Bruce Trombecky)

◂ An A-7B of VA-215, while assigned to USS *Oriskany*. VA-215 was one of the first squadrons to take the A-7B into battle, making the first combat cruise aboard carrier USS *Enterprise* (CVN-65) in 1969. (Bruce Trombecky)

◂ An A-7D of the 57th Fighter Weapons Wing at Nellis AFB, Nevada in 1975. A total of 459 A-7Ds were built, equipping three TAC wings and two Air National Guard squadrons. (Don Logan)

▾ A-7Ds eventually were painted in a variety of 'toned-down' camouflage schemes, depending on the ANG unit to which they belonged. The open access panel of this Corsair II reveals both the built-in electronics test panel and avionics 'black boxes.' (Andre Jans)

▲ A TA-7C of VA-122. After demonstrating the two-seat Corsair II at various naval air stations, Vought was awarded a contract to modify sixty TF30-powered Corsair IIs (twenty-four A-7Bs and thirty-six A-7Cs) into two-seat trainers, to be designated TA-7C. The first converted TA-7C made its first flight on 17 December 1976 and was delivered to the Navy on 31 January 1977. The TA-7Cs were delivered to VA-122 and to VA-174. (Andre Jans)

▼ An A-7D on takeoff. Main gear doors are just completing the retraction cycle. The pod carried on the outboard station is commonly referred to as a 'travel pod.' It is used to carry the pilot's luggage and/or any required specialized support equipment when the aircraft travels to a display venue. (Andre Jans)

▲ The fairing at the base of the rudder on the A-7E contains the AN/ALQ-100 or AN/ALQ-126 ECM antenna.

▸ Underside view of a typical A-7. All versions are equipped with an arresting hook, which is retracted hydraulically using PC 2 hydraulic system pressure and extended by a combination of gravity loads and hydraulic pressure. When the hook is extended, downward force is maintained by accumulator pressure acting against the actuator.

◂ The prototype TA-7C was equipped with a drag chute at the base of the vertical fin. This was to be used for spin recovery during the test phase of the modified A-7E. (Vought)

▾ The LTV prototype TA-7C on its first flight. The prototype TA-7C was originally designated YA-7H, then YA-7E, before the TA-7C designation was adopted. This aircraft became the company test and demonstration vehicle for hundreds of A-7 modifications and new systems applications. It operated for thirteen years in that capacity before ultimately being sold for scrap. (Vought)

▲ An A-7E of VA-105 on the ramp at NAS Cecil Field, Jacksonville, Florida. The large screen affixed to the intake prevents engine damage due to ingestion of foreign objects and protects ground personnel during ground run-ups of the engine. Ramp areas of Master Jet bases are equipped with built-in, yellow/black striped, air start units. (Bruce Trombecky)

▼ An A-7B-2-CV of VA-87. It has been modified with the addition of the ECM antenna at the base of the tail. VA-87 A-7Bs were assigned to CVW-6 aboard USS *Franklin D. Roosevelt* from 1970 to 1975. (Bruce Trombecky)

▼ An A-7E of VA-22. VA-22 operated the A-7E from 1971 to 1990 with assignments to CVW-15 aboard USS *Coral Sea* (CVA 43) and USS *Kitty Hawk* (CVA 63), and CVW-11 aboard USS *Enterprise* (CVN-65). 'NL' tail code indicates assignment to *Coral Sea* for this 'Fighting Redcocks' Corsair II. (Bruce Trombecky)

▲ Engine cranking is accomplished by the air turbine starter which is driven by an external source of pressurized air. The air hose is connected to a fuselage nipple fitting which directs pressurized air to the starter air inlet flange. (Lou Drendel)

▸ The A-7's rudder becomes effective at 50 to 75 knots indicated air speed (KIAS), depending on gross weight of the aircraft. Prior to that speed, nosewheel steering is used for directional control. (Lou Drendel)

▲ The open access panel on this A-7E of VA-105 reveals the autopilot, trim, and rudder feel units and controller. The fin cap containing the UHF and IFF antennas has been removed.

▲ An A-7A of VA-87. The fairing at the top of the rudder contains the antenna for the AN/ALQ-100/51A receiver (rear) and the AN/APR-25(V) receiver (front). (Dr. J.G. Handelman)

◂ An A-7E of VA-174. The fairing above the A-7E's rudder was enlarged to house enhanced ECM receivers. This also resulted in relocation of the tail light further aft on the fairing. (Dr. J.G. Handelman)

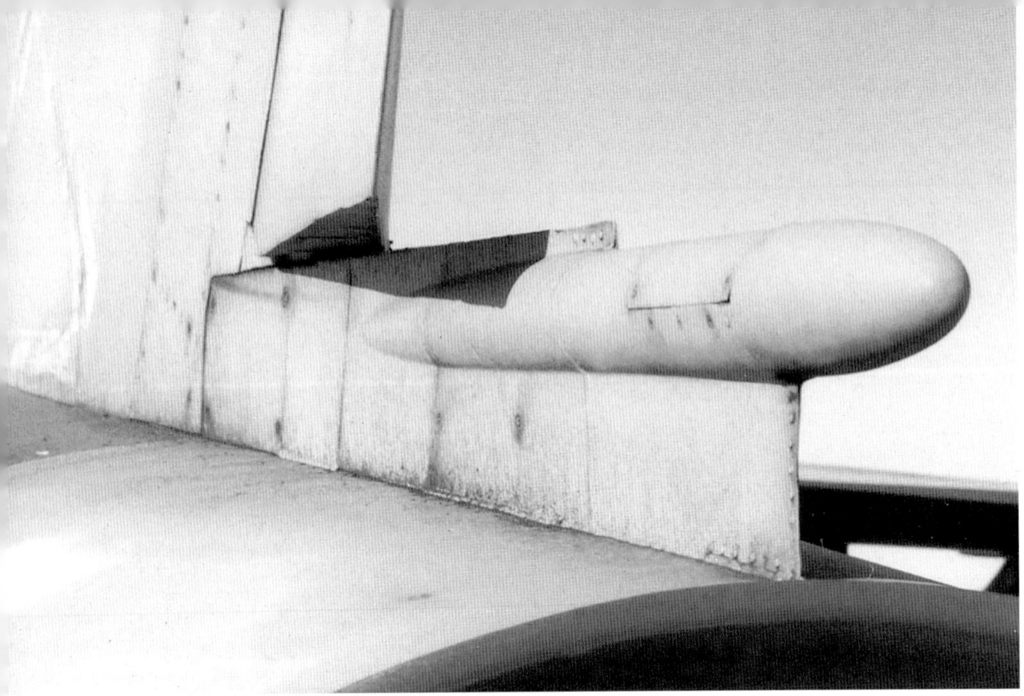

▲ The fairing for one of the ECM antennas. The AN/ALQ-126 installation typically uses forward and aft antennas with 60-degree beam width and 15-degree depression to cover surface threats.

▲ The A-7E has carried several defensive countermeasure systems, including the ALR-45 radar warning receiver and ALR-50 surface-to-air missile warning system (Magnavox), the ALQ-126 electronic countermeasure system (Sanders), and the APR-43 tactical radar warning system (Loral).

◄ The all-moving horizontal stabilizer is 93.75 square feet in area and is mounted with 5 degrees of dihedral (tips up). The sweep is 45 degrees at ¼ chord. The control stick and rudder pedals operate mechanical linkages to position servo valves of hydraulic power control cylinders. The power control cylinders are mechanically linked to the control surfaces and cause movement of the selected surface. There is no air load feedback to the stick or rudder pedals, so a series of springs provide artificial feedback feel to the pilot.

▲ A-7As of VA-86 'Sidewinders' served aboard USS *Coral Sea* (CVA 43) with CVW-15 from September 1969 to July 1970. It was the fifth combat cruise of the Vietnam War for *Coral Sea*.

▲ Aft end of the TF-41-A-2 engine in the A-7E, showing the low pressure compressor section.

▸ Tail pipe of the A-7E. The exhaust system consists of a reinforced, tapered, sheet metal tailpipe bolted to the rear flange of the exhaust bypass mixer. The duct converges in diameter from the exhaust bypass mixer flange to the jet nozzle opening.

▲ The 'River Rattlers' of Reserve Attack Squadron 204 (VA-204) received their A-7Bs on March 15, 1978. They were assigned to CVWR-20, and flew the A-7 until 1991, when they traded their Corsair IIs for F-18 Hornets.

◀ The final operational combat markings for the A-7 were applied to Corsair IIs of VA-46 and VA-72 aboard USS *John F. Kennedy* during Operation Desert Storm in January/February of 1991. The small red light is a formation light, used in low-visibility situations or at night for station-keeping.

▼ On 1 February 1968, the 'Golden Warriors' were established as Attack Squadron 87 at NAS Cecil Field, Florida. The squadron flew its first combat mission on 4 March 1969 from USS *Ticonderoga* (CVA 14), striking enemy targets in South Vietnam. In August 1970, VA 87 joined Air Wing 6 and made twelve deployments aboard three different carriers: USS *Franklin D. Roosevelt* (CV 42), USS *America* (CV 66) and USS *Independence* (CV 62). (Dr. J.G. Handelman)

▲ An A-7D of the 174th TFS/185th TFG, Sioux City, Iowa, lands at Nellis AFB, Nevada, during a Red Flag exercise. It is carrying a pod which records data for the range exercises. (Ted Carlson)

▼ A-7D-3-CV Corsair II at Davis Monthan AFB, Arizona. Although this is not an early A-7D, it lacks the in-flight refueling receptacle common to most D-models. The black outline on the vertical tail is the Loran antenna. (Don Logan)

▲ The ultimate in wartime paint. A CAG A-7E with no distinctive markings aboard USS *America* (CV 66) in 1985. (Lou Drendel)

▲ The starboard rear fuselage of an A-7E. The area under the stabilator is bare metal. (Lou Drendel)

▼ An A-7D-4-CV Corsair II at Davis Monthan AFB, Arizona, in overall Gunship Gray (FS595a: 36118) camouflage with a multiple ejector rack (MER) on the center pylon. The MER is a weapon suspension unit that attaches to an aircraft's main racks (pylons) and which can carry up to six weapons, such as the Mk 20, Mk 82 'Slick,' Mk 82 'Snakeye,' CBU-87, or CBU-97. The Air Force stopped using MERs in the early 1990s and currently uses only triple ejector racks (TERs).

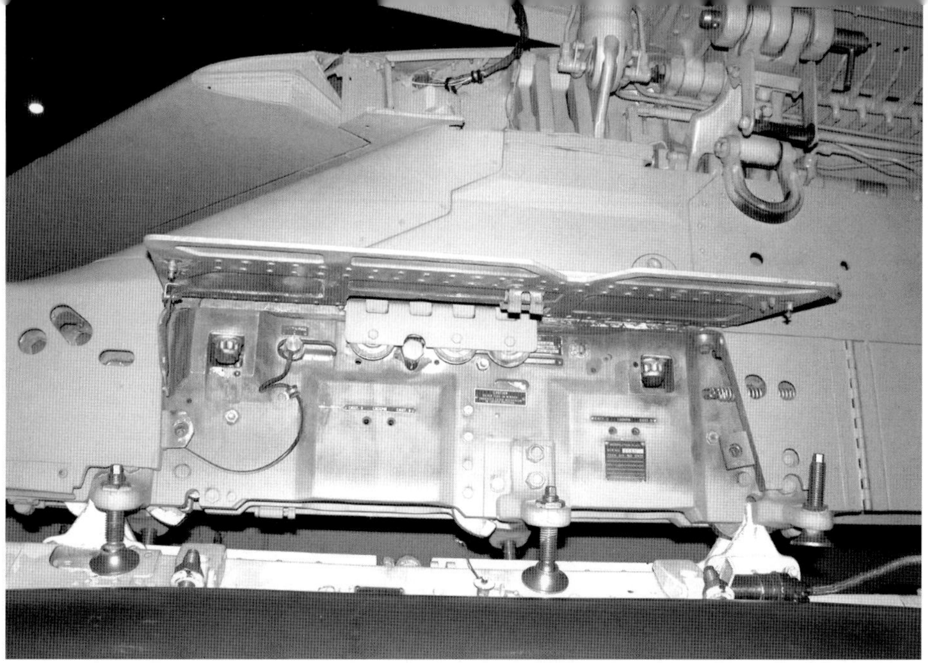

▲ External stores can be mounted on six wing pylons and two fuselage pylons. The actual number of stores that can be carried on any one pylon depends on the load being carried on other pylons, on whether the stores are to be released in normal release sequence, and on aircraft gross weight and center of gravity (CG) considerations. (Lou Drendel)

▶ Port inboard wing pylon and fuselage pylon. The fuselage pylon is loaded with an AIM-9 Sidewinder air-to-air missile. Fuselage pylons have launch priority over wing pylons if they are selected. (Lou Drendel)

▼ A-7E inboard pylon in clean configuration. (Lou Drendel)

Starboard outboard pylon and wing fold mechanism. The normal release sequence is from wing pylons 1, 8, 2, 7, 3, and 6 (numbered from outboard to inboard, left to right) if all wing stations are selected. If stores are released in the order of normal priority, excessive asymmetric load conditions are not developed. (Lou Drendel)

▲ Wing outer panels are folded, or spread and locked, using PC 2 system hydraulic pressure. Folding and spreading is accomplished by a single hydraulic actuator for each panel. This actuator exerts sufficient force to overcome airloads imposed by winds up to 40 knots at 90 degrees to the fuselage center line during fold operation, or 36 knots during spread operation. Locking is accomplished by hydraulic hinge pin cylinders (two per panel) which insert hinge pins through joints formed by hinge lugs when the panels are spread. Mechanical lock latches are provided for safing the hinge pins in place while parked, and warning flags to visually indicate latch position. (Lou Drendel)

▶ The ailerons on the A-7 provide roll control. Rolls are accompanied by an adverse yaw tendency at low speeds, which decreases as speed builds. There is an aileron-rudder interconnect to aid in coordinated rolls. When the interconnect is inoperative, the adverse yaw tendency is greater. (Lou Drendel)

▼ The multi-position flaps on the A-7 require significant trim changes to overcome stick pressures, and the flaps produce airframe buffet at more than 30 degrees of deflection. Full flaps extension and/or retraction takes ten to eleven seconds. (Lou Drendel)

▾ The fuel dump is located on the trailing edge of the wing, between the aileron and flap. Dump rate begins at 300 gallons per minute and decreases as flow progresses. Higher rates can be maintained by increasing tank pressurization with high engine rpm. (Lou Drendel)

▴ Six pylons are provided for stores carriage on the wings. From left to right, the pylons are numbered 1, 2, and 3 (left wing) and 6, 7, and 8 (right wing). Racks 4 and 5 are on the fuselage. A parent rack is installed on each pylon. All stores or additional racks attach to the parent racks. (Lou Drendel)

▾ Aft end of the Aero 1-D 300-gallon external tank mounted on the inboard pylon. External tanks can be mounted on stations 1, 3, 6, and/or 8. (Lou Drendel)

◂ One parent rack is installed on each wing pylon. A single store, missile launcher, or a multiple carriage rack is supported by hooks from each parent rack. During release of a single store, ejector cartridges open the hooks to release the store and cause an ejector foot to kick the store downward. If the ejector cartridge misfires during jettisoning, the store is gravity-dropped.

◂ An A-7 in Vietnam-era paint scheme of flat Light Gull Gray (FS 595a: 36440) upper surfaces and gloss White (FS 595a: 17875) undersurfaces and control surfaces (ailerons, flaps, stabilators, and rudder). Fuselage pylons have an LAU-7/A launcher installed for carriage of AIM-9 Sidewinder missiles.

◂ Attack Squadron 147 was commissioned as the Navy's first A-7A Corsair II squadron on 1 February 1967 at NAS Lemoore, California, and was the first squadron to employ the Corsair II during the Vietnam War while deployed to the Western Pacific aboard USS *Ranger* (CVA 61). The squadron completed five combat deployments while conducting air operations over Vietnam. Throughout the Argonauts' Corsair II years, the squadron won the coveted Battle 'E' in 1977 as the top Corsair squadron in the Pacific Fleet, and also received three CNO Safety Awards and a Meritorious Unit Commendation for bombing excellence. The A-7E was retired following a Western Pacific-Indian Ocean deployment in February 1989, and Attack Squadron 147 was officially redesignated as Strike Fighter Squadron 147 (VFA 147) on 20 July, 1989, trading in their Corsair IIs for F/A-18 Hornets.

▾ An A-7E of the 'Royal Maces' turns final at NAS Miramar. Attack Squadron 27 (VA-27) was established on 1 September 1967 and was the second A-7 squadron to deploy to Vietnam. Redesignated Strike Fighter Squadron 27 (VFA-27) on 24 January 1991, VA-27 was the last A-7 squadron at NAS Lemoore to transition to the Hornet. (Ted Carlson)

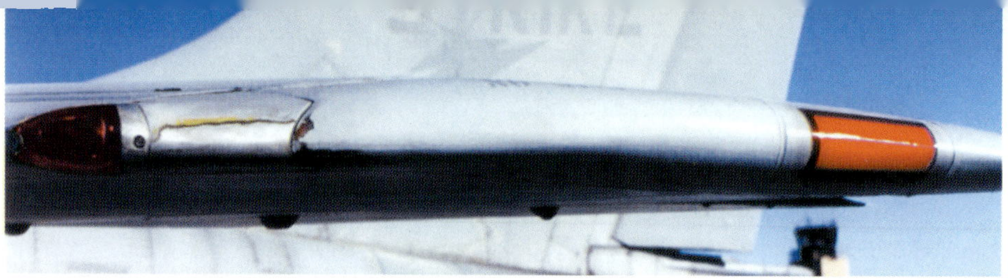

▲ Formation lights are located on each wingtip. They are normally used for formation flight in low-light conditions that do not allow use of normal anti-collision lighting for tactical reasons. (Dave Mason)

▶ Port aileron on the A-7D in full deflection. Ailerons can be deflected 16 degrees up or down with control stick movement, but only 14 or 13 degrees, respectively, with trim input. A maximum deflection of 25 degrees can be achieved through the use of stick, rudder, trim and Automatic Flight Control System (AFCS). (Dave Mason)

▼ An A-7B of VA-46 crosses the ramp of USS *John F. Kennedy.* It is carrying an inert Bullpup missile on station 8, a Bullpup launcher on station 1, and TERs on stations 2 and 7. VA-46 transitioned to the Corsair II in 1968 and was one of the last two operational regular Navy A-7E squadrons.

▲ The dual nosewheels are independently mounted on a common axle which is attached to a conventional air-oil shock strut. Catapult provisions are built into the nose gear. A mechanical linkage will center the nose gear during retraction if the wheels are off-center. The nose gear retracts aft. A tension spring arrangement on the nose gear drag strut aids nose gear extension if emergency hydraulic pressure is low. (Lou Drendel)

◄ Nose gear steering on the A-7 is electrically controlled, hydraulically actuated and provides power steering via movement of the rudder pedals. The flaps must be down 20 degrees or more to ensure full nose gear steering. With landing gear extended, hook extension causes the nose gear to hydraulically center. (Lou Drendel)

Catapult Hook-up

Procedure:

1. Install catapult holdback tension bar.
2. Lower launch bar and taxi airplane into catapult.

NOTE

Launch bar and holdback bar will automatically hook up to the catapult.

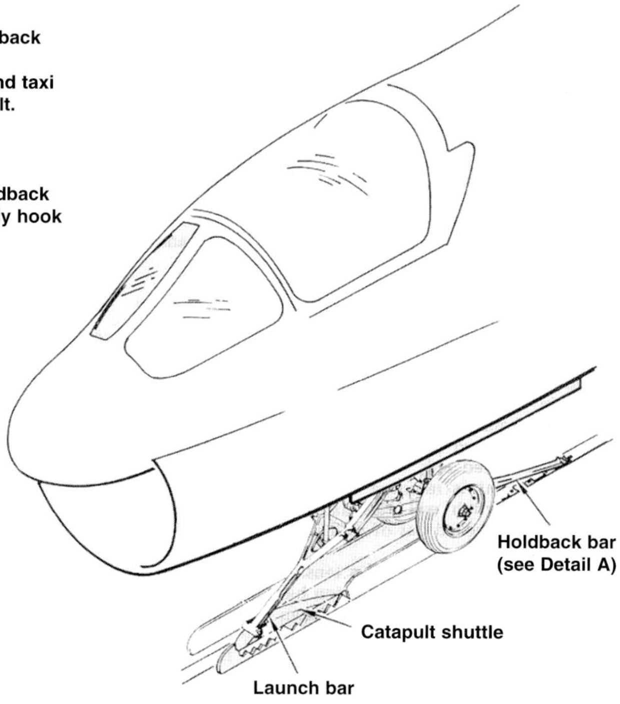

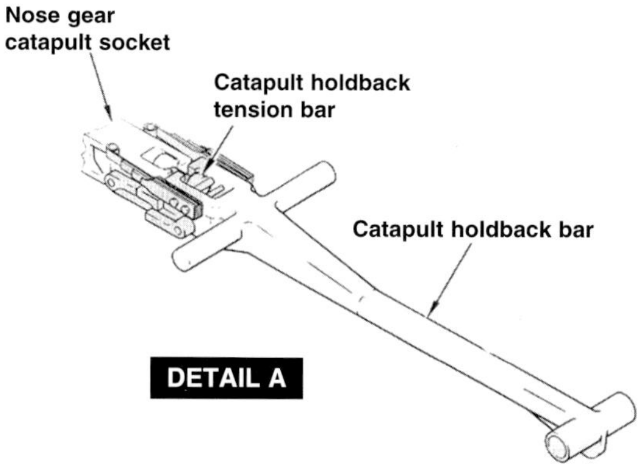

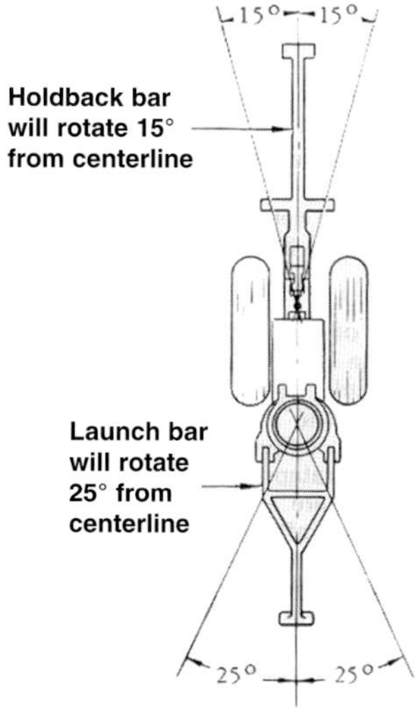

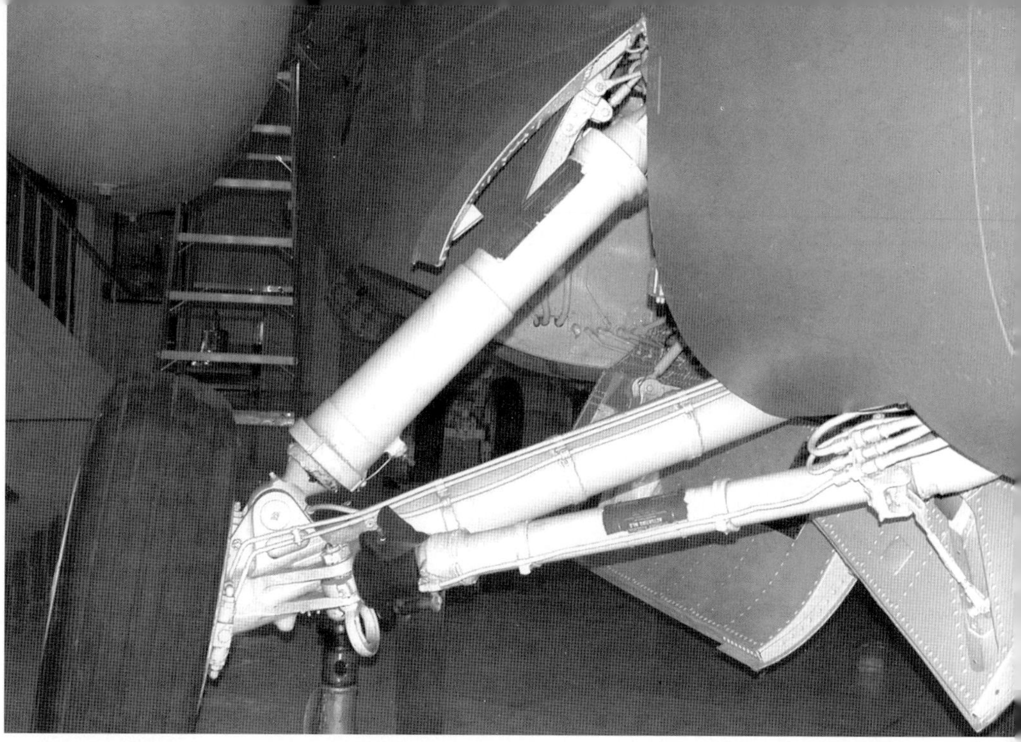

▲ Each main gear is a tripod arrangement consisting of an air-oil shock strut, a drag strut, and a tension strut. Gear doors completely enclose the retracted landing gear. This is the port main landing gear, looking forward. (Lou Drendel)

▼ The port A-7E main gear wheel well, viewed from the front. (Lou Drendel)

◂ The catapult launch bar extends forward from the landing gear. It is lowered to engage the catapult shuttle on the carrier deck. As the aircraft is moved forward into position on the catapult, the holdback assembly will be installed. When the launch bar drops over the shuttle, the aircraft will be stopped. The holdback assembly is engaged. At light fuel weights, the pilot is required to turn up the engine in order to depress the nose oleo so as to allow the launch bar to drop over the shuttle. (Dave Mason)

▾ The approach lights are located on the port nosewheel door. The lights are on when the landing gear handle is in 'WHLS DOWN' position without weight on the right main gear. The lights are on steady when the hook switch is in either 'BYPAS' or 'CARRIER' position with the arresting hook down. The lights will flash if the hook switch is in 'CARRIER' with the arresting hook up. (Lou Drendel)

▾ The starboard main gear. Labels on the strut are instructions for servicing the air-oleo strut. (Dave Mason)

▲ On 1 November 1967, VA-105 was recommissioned at Cecil Field, flying the new A-7A Corsair II. The 'Gunslingers' embarked on their first deployment to Southeast Asia, participating in combat operations in the Gulf of Tonkin from January to October 1969. On 10 January 1991, VA-105 was redesignated VFA-105 as the squadron transitioned from the A-7E to the F/A-18C. A-7E on final approach carries an Air Combat Maneuvering Instrumentation (ACMI) pod on the port fuselage station. (Ted Carlson)

◂ An A-7A of VA-147 approaches the ramp of USS *Ranger* during the initial combat cruise of the Corsair II.

▸ An A-7B comes in close for a carrier landing. The A-7's landing gear extends significantly to cushion the shock of landing on deck. Main wheels and tires are size 28 x 9-12; nosewheels and tires are size 22 x 5.50.

▾ Factory view of an A-7E fuselage under construction with the weight off the main gear as it would be in flight. Power-boosted, tri-metallic, self-adjusting disk brakes are mounted on each main wheel. Brake pressure is normally supplied from the PC 2 hydraulic system through power-boosted brake cylinders controlled by pressure on the tops of the rudder pedals (commonly referred to as 'toe brakes'). (Lou Drendel)

▾ Port main gear well, looking aft. Weight on the main gear compresses the oleo shock strut completely, resulting in movement of the wheels from positive to negative camber. (Lou Drendel)

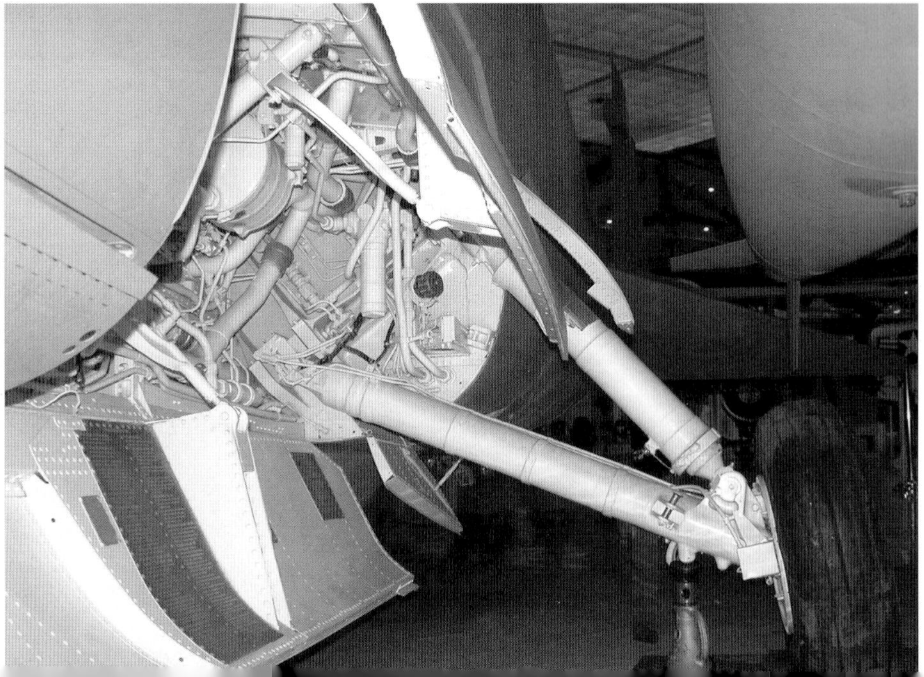

The starboard main gear of an A-7E. Landing gear doors require up to eight seconds to close and lock. Any time the flap handle is placed in the 'ISO UTILITY' position less than eight seconds after all landing gears indicate 'UP,' the main landing gear will extend. This extension may not always occur immediately but can occur later in the flight as airspeed and/or normal accelerations increase. (Lou Drendel)

▲ A-7E aboard USS *America*, manned up and ready for engine start prior to commencing carrier qualifications in 1985. (Lou Drendel)

◂ A-7D interior lighting. Prominent features include the gun sight/head-up display (HUD) on top of the instrument panel glare shield and the large yellow ejection seat handle above and behind the pilot's head.

▲ Canopy opening handle in the released position. The handle is flush-mounted but springs out when a handle latch is released. Turning the handle counterclockwise unlocks the canopy, and turning the handle clockwise to a horizontal position locks the canopy. (Dave Mason)

▲ The one-piece, clamshell type canopy is attached to the airframe by two pivot bolts immediately aft of the ejection seat. Three rear-view mirrors are mounted on the forward canopy frame. The canopy is normally manually closed and is locked when four locking hooks are engaged in four rollers in the canopy frame. (Lou Drendel)

▼ YA-7D 67-14583 at Edwards AFB. It lacks the refueling receptacle and is equipped with a test instrumentation boom on the nose. (Peter Steendam)

Pilot's Equipment

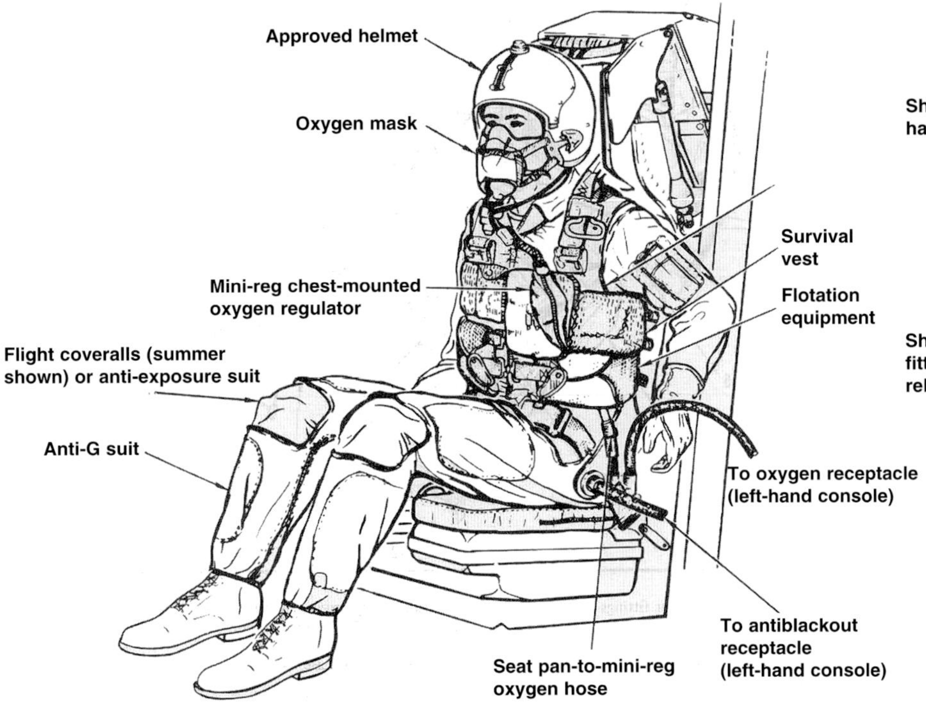

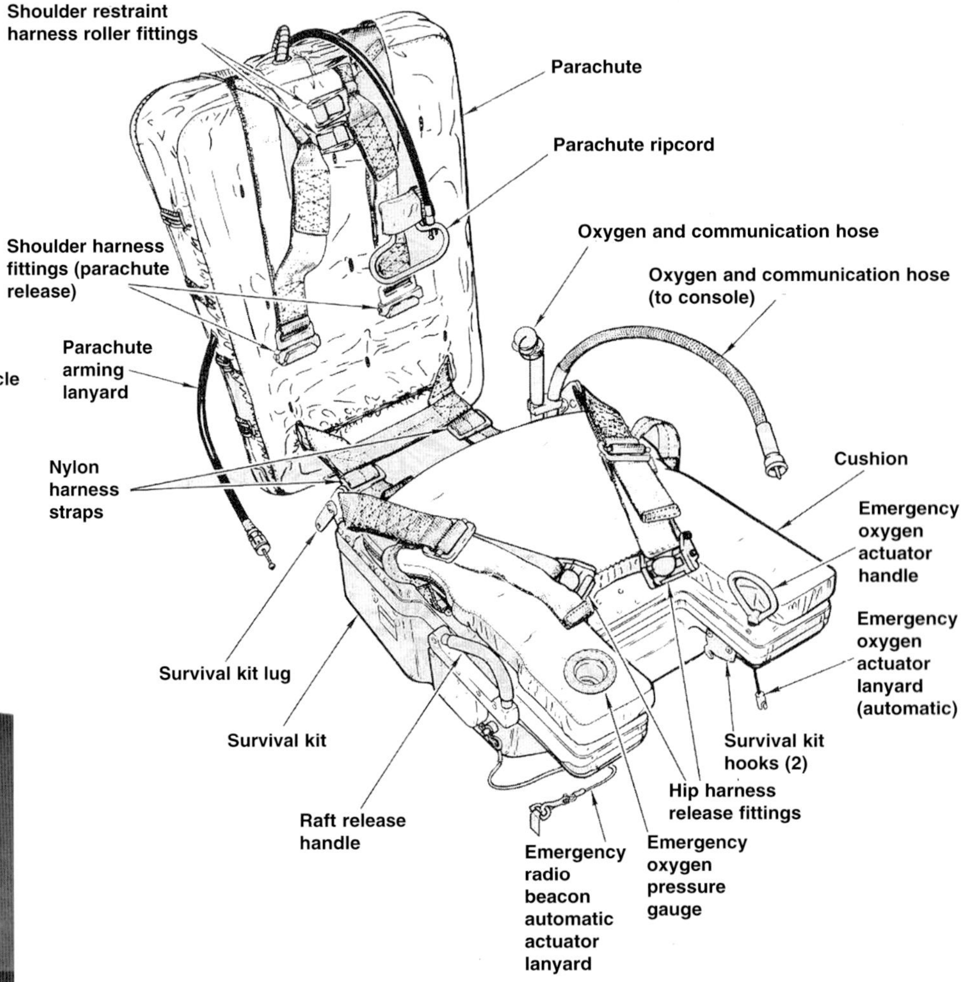

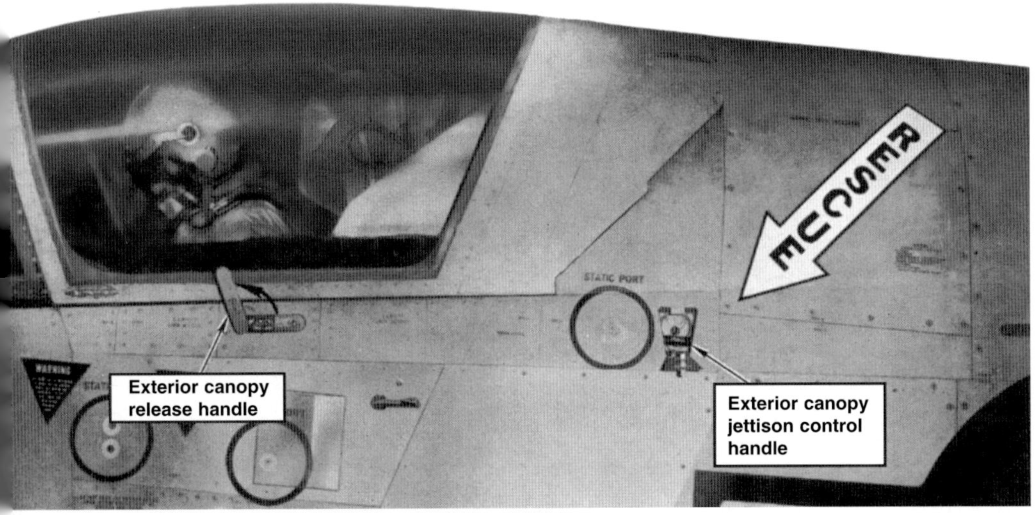

◀ **The canopy can be jettisoned for landing emergencies, ditching, and ground rescue and is automatically jettisoned during the ejection sequence. Pulling the Emergency Canopy Jettison Handle above the forward left console or either exterior jettison handles fires an initiator connected to the pulled handle.**

ESCAPAC 1-G2 Ejection Seat (A-7A/B)

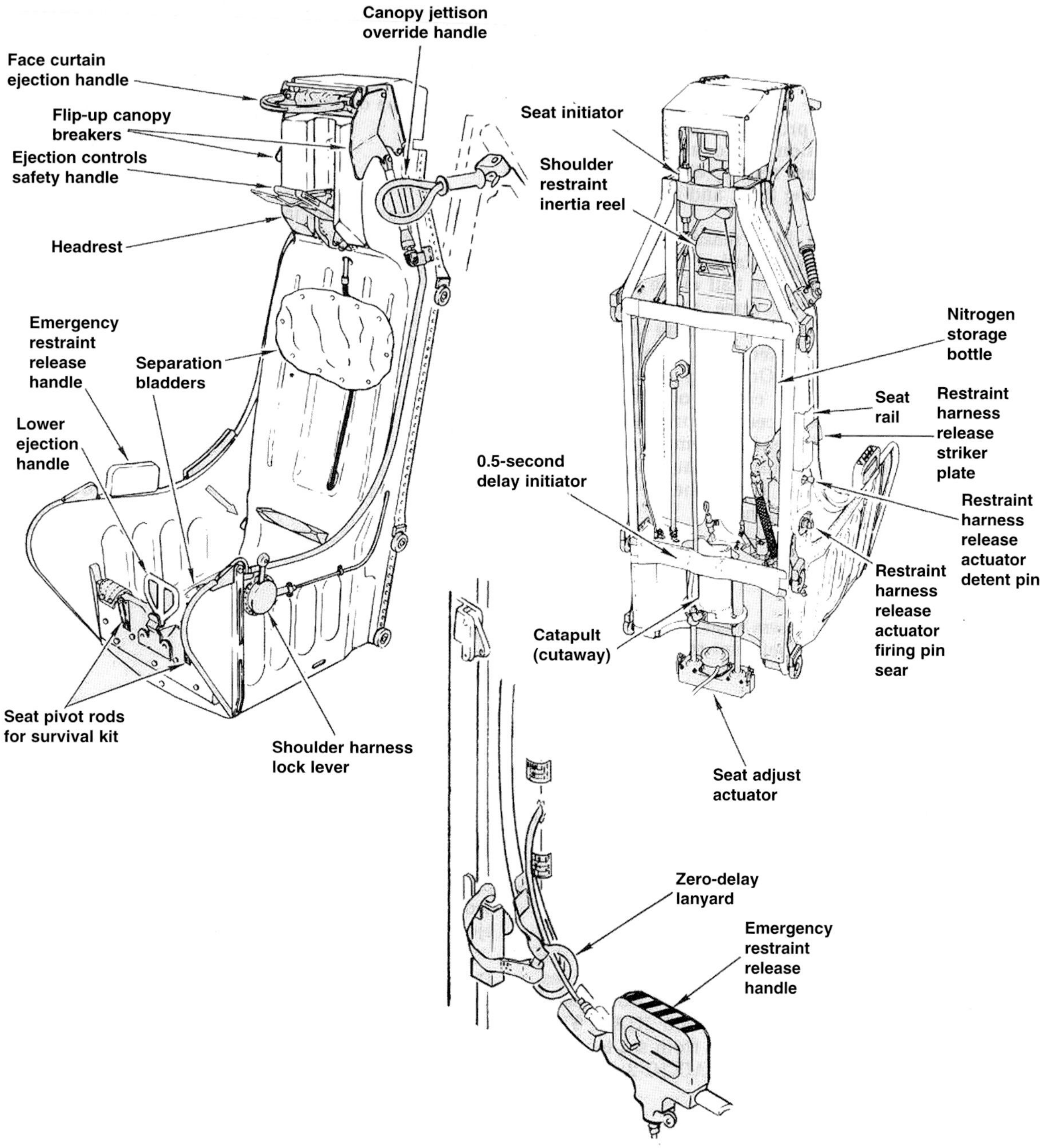

Gunsight/HUD

A-7A/B Gunsight/HUD

A-7E Gunsight/HUD

Cockpit of the A-7P. Though converted from the A-7A, the A-7P was equipped with A-7E avionics. The two round screens are radar (top) and a moving map display (bottom).

▲ EA-7L instrument panel.

▼ Cutaway of the TF-41 engine. The rear of the engine is at the left, with the turbine exhaust cone prominent. The A-7C was equipped with the Pratt & Whitney TF-30-P-408 engine. The A-7E was equipped with the Allison TF-41-A-2 turbofan. (Lou Drendel)

▲ Details of the AN/AVQ-7(V) Head-Up Display (HUD). The HUD is an optical and electronic device that projects attack and flight information in symbolic form into the pilot's field of view on a combiner glass similar to a standard gunsight.

▼ The combiner glass has two positions, fore and aft. The forward position is used for en route navigation, and the aft position is used during attack and landing. A standby reticle is provided as a backup in the event of HUD failure.

A-7D Cockpit

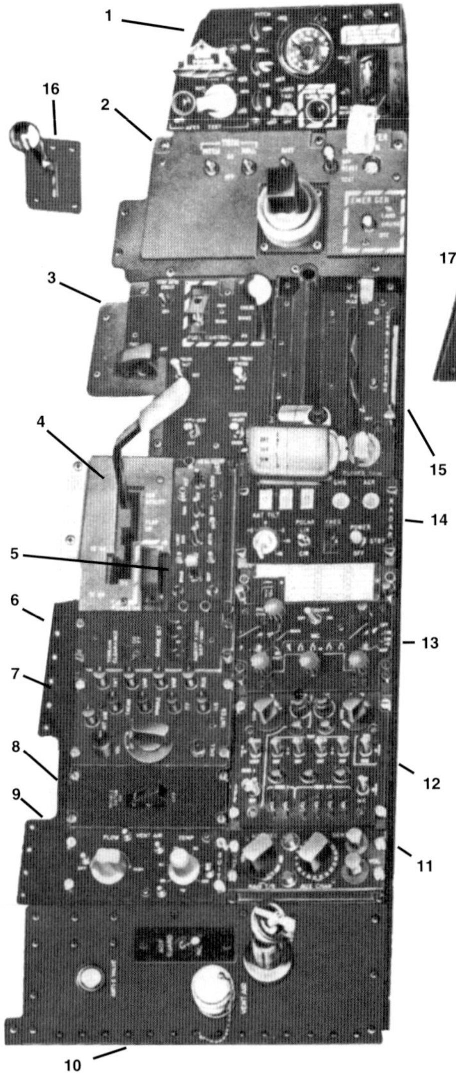

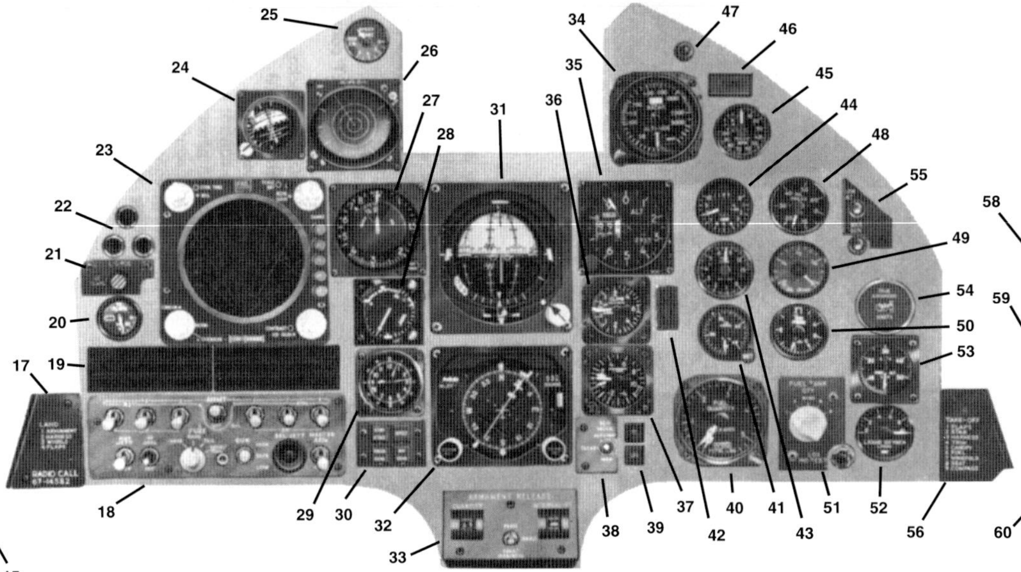

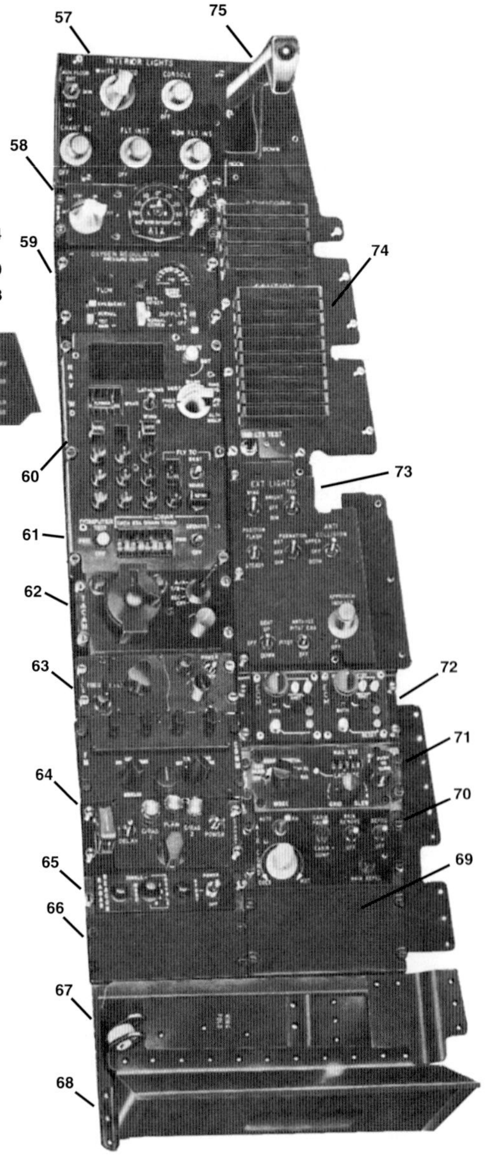

1. Landing gear system controls; emergency power pitch, roll, and yaw trim indicators
2. Generator selector panel
3. Fuel and emergency brake panel
4. Flap control
5. AFCS control panel
6. Terrain clearance/range set control
7. Communications set controls
8. Radar homing/warning controls
9. Suit temperature controls
10. Pilot services panel
11. ADF control panel
12. IFF control panel
13. UHF radio control panel
14. Radar control panel
15. Throttle, master fuel, rudder trim controls
16. Canopy jettison (on left longeron)
17. Landing checklist
18. Armament selection panel
19. Armament advisory light panel
20. Flap position indicator
21. Leading edge flaps indicator
22. Landing gear position lights
23. Forward-looking radar
24. Standby attitude indicator
25. Speed brake indicator
26. APR-36 threat analyzer
27. Mach-airspeed indicator
28. Angle-of-attack indicator
29. Clock
30. Attack master function switches (4)
31. Attitude director indicator
32. Horizontal situation indicator
33. Armament release panel
34. Radar altimeter
35. Barometric altimeter
36. Vertical velocity indicator
37. Accelerometer
38. Heading mode switch
39. Terrain following/landing switches
40. Fuel quantity indicator
41. Turbine outlet pressure indicator
42. Master caution light
43. Fuel flow indicator
44. Turbine outlet temperature indicator
45. Tachometer
46. Fire warning light
47. Fire warning switch
48. Oil pressure indicator
49. Oil quantity indicator
50. Hydraulic pressure indicator
51. Fuel tank quantity selector
52. Liquid oxygen indicator
53. Cockpit pressure altimeter
54. True airspeed indicator
55. Marker beacon/low altitude lights
56. Takeoff checklist
57. Interior lights control panel
58. Doppler radar controls
59. Oxygen regulator
60. Navigation/wind data computer
61. TACAN control panel
62. ILS control panel
63. VHF FM control panel
64. Speech security control panel
65. Radar beacon control panel
66. Blank panel
67. Wing fold controls
68. Map case
69. Blank panel
70. Air conditioning control panel
71. Inertial measurement set controls
72. ECM pod control panel
73. Exterior lights controls
74. Advisory/caution light panels
75. Arresting hook handle

A-7E Cockpit

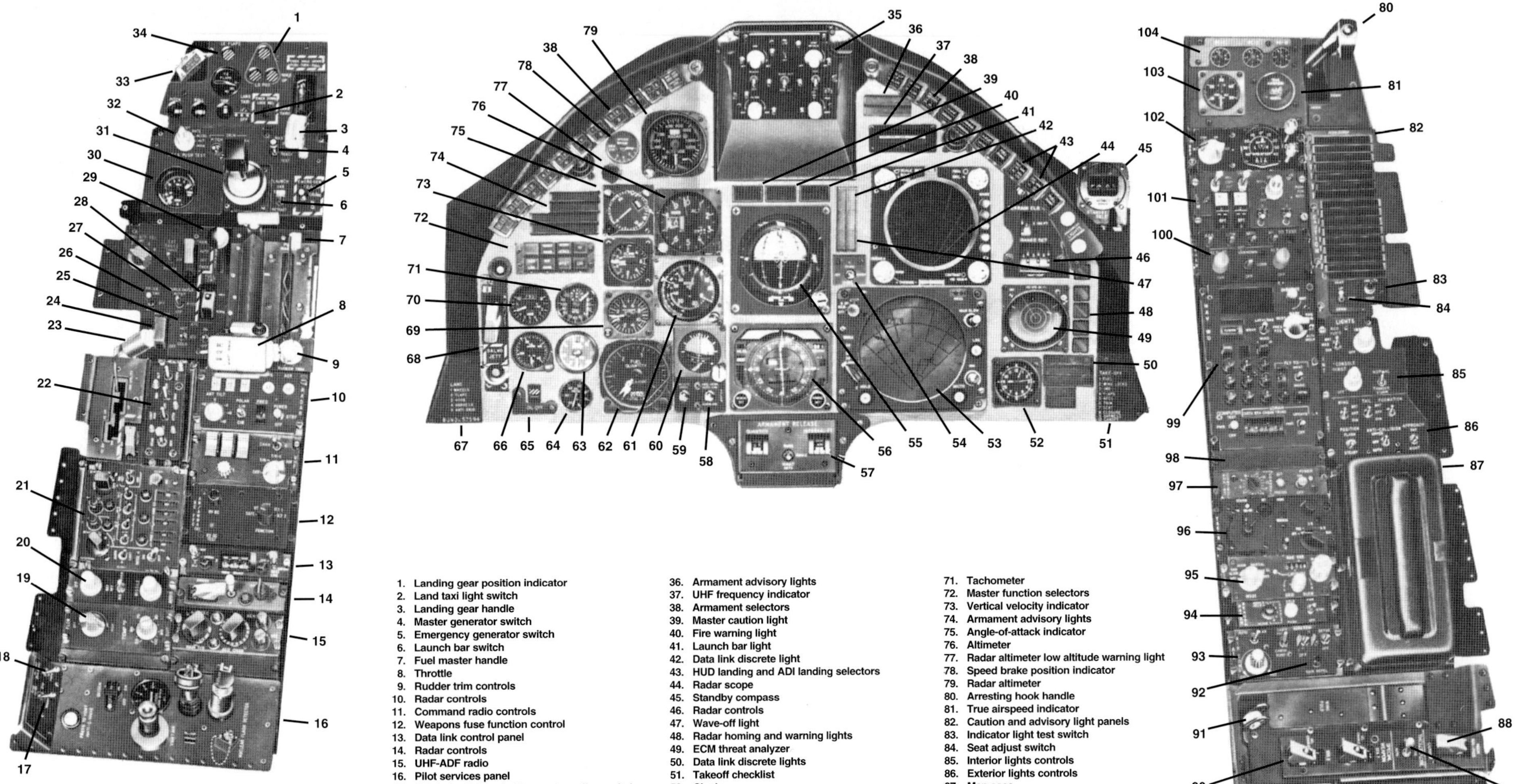

1. Landing gear position indicator
2. Land taxi light switch
3. Landing gear handle
4. Master generator switch
5. Emergency generator switch
6. Launch bar switch
7. Fuel master handle
8. Throttle
9. Rudder trim controls
10. Radar controls
11. Command radio controls
12. Weapons fuse function control
13. Data link control panel
14. Radar controls
15. UHF-ADF radio
16. Pilot services panel
17. Ship's inertial navigation system align switch
18. ADF guard switch
19. Suit temperature control
20. Audio control
21. IFF controls
22. Automatic flight control system switches
23. Flap handle
24. Anti-skid switch
25. Approach power control system panel
26. Probe switch
27. Double datum Pitot switch
28. Fuel control switch
29. Emergency brake handle
30. Pitch and roll trim indicator
31. Slew control
32. AFCS push-to-test
33. Emergency power handle
34. Flap indicators
35. Head-up display controls
36. Armament advisory lights
37. UHF frequency indicator
38. Armament selectors
39. Master caution light
40. Fire warning light
41. Launch bar light
42. Data link discrete light
43. HUD landing and ADI landing selectors
44. Radar scope
45. Standby compass
46. Radar controls
47. Wave-off light
48. Radar homing and warning lights
49. ECM threat analyzer
50. Data link discrete lights
51. Takeoff checklist
52. Clock
53. Projected map display
54. Shrike switch
55. Attitude director indicator
56. Horizontal situation indicator
57. Armament release panel
58. Heading mode switch
59. Walleye switch
60. Standby attitude indicator
61. Airspeed indicator
62. Fuel quantity indicator
63. Fuel flow indicator
64. Oil pressure indicator
65. Oil quantity indicator
66. Turbine outlet pressure indicator
67. Landing checklist
68. Jettison buttons and switch
69. Accelerometer
70. Turbine outlet temperature
71. Tachometer
72. Master function selectors
73. Vertical velocity indicator
74. Armament advisory lights
75. Angle-of-attack indicator
76. Altimeter
77. Radar altimeter low altitude warning light
78. Speed brake position indicator
79. Radar altimeter
80. Arresting hook handle
81. True airspeed indicator
82. Caution and advisory light panels
83. Indicator light test switch
84. Seat adjust switch
85. Interior lights controls
86. Exterior lights controls
87. Map case
88. Hook bypass switch
89. Thermal closure switch
90. Special weapons switches
91. Wing fold controls
92. Rain repellent button
93. Air conditioning and pressurization panel
94. Radar beacon controls
95. Inertial measurement set controls
96. TACAN controls
97. Automatic carrier landing system controls
98. Blank panel
99. Tactical computer control panel
100. Radar homing and warning controls
101. ECM controls
102. Doppler controls
103. Cabin pressure altimeter
104. Hydraulic pressure indicators

▲ An A-7B of VA-72 at NAS Cecil Field, Florida, 1973.

◄ The control stick contains the designate switch, release enable switch, and the trigger switch. The designate switch is used to designate and un-designate targets in attack modes and update modes and also controls caging of Walleye missiles. The crank handle in front of the base of the control stick is used to adjust fore-and-aft position of the rudder pedals. (Lou Drendel)

▼ A single Pitot tube, mounted just aft of the radome on the left side of the aircraft, supplies impact pressure to the Mach and airspeed indicator and to the air data computer. The Pitot tube is heated by an electrical element to prevent or eliminate ice build-up in the open end. (Dave Mason)

A Portuguese A-7P pilot awaits command to taxi. The angle-of-attack indicator in the cockpit receives signals from a transducer located on the left side of the fuselage (aft of the ejection seat triangle). The indicator reflects indicated angle of attack (true angle is less than indicated). Rescue arrows point to the port emergency canopy jettison handle. There is also a handle on the starboard side in the same position. This pilot has stowed his maps on the glareshield. (Andre Jans)

Servicing Diagram

FLUID SPECIFICATION		
	USAF	NATO
Fuel	Normal: MIL-T-5624 (Grade JP-4)	F-40
Engine Oil	MIL-L-7808	O-148
Hydraulic Fluid	MIL-H-5606B	H-515
Nitrogen	BB-N-411b	
Liquid Oxygen	MIL-O-27210 (Grade A, Type I or II)	None
Turbine Oil	MIL-L-6085	O-147

SYSTEM SERVICING	
Engine Constant Speed Drive Jet Fuel Starter	Engine Oil
PC 1 PC 2 PC 3 Viscous Dampers	Hydraulic Fluid
Liquid Oxygen Converter	Liquid Oxygen
All External and Internal Fuel Tanks	Fuel
All Hydraulic Accumulators	Nitrogen
Air Conditioning Turbine	Turbine Oil

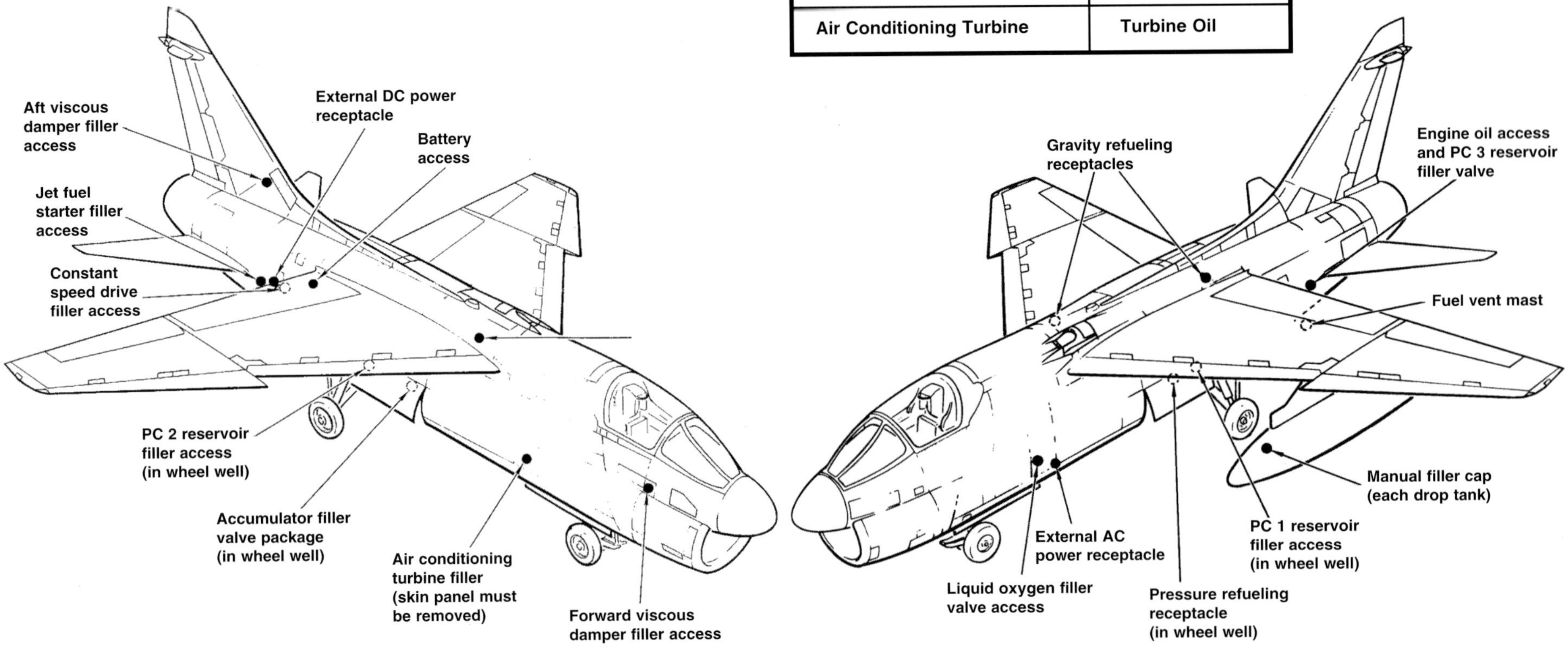

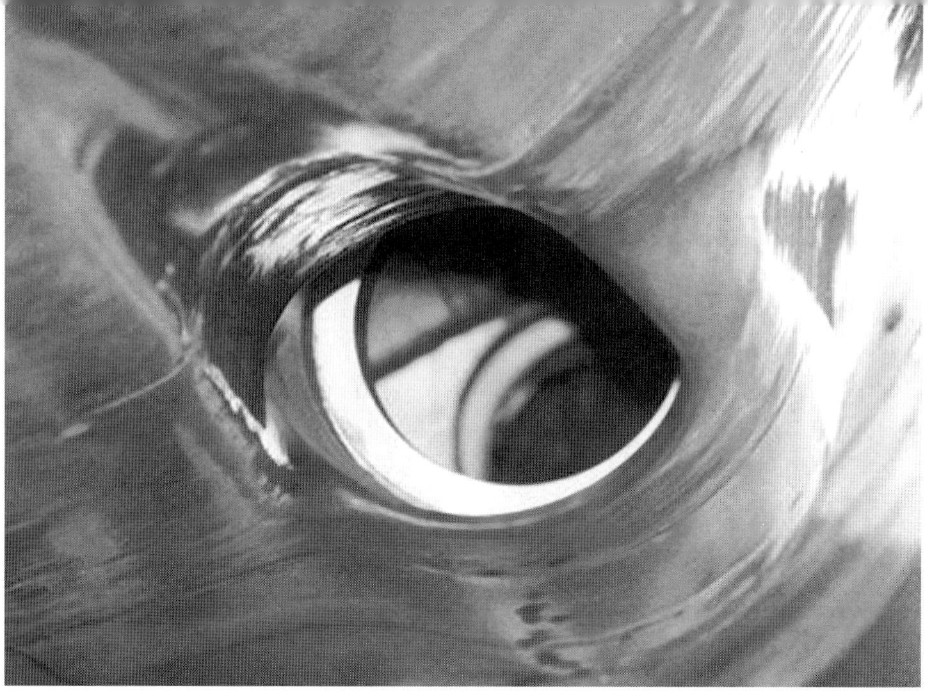

▲ The business end of the M61A1 cannon in an A-7E. The M61A1 gun is electrically controlled and hydraulically driven and is fed by a linkless feed system. The gun consists of a rotating cluster of six barrels, which fires 20 mm electrically primed ammunition at two rates, high (6,000 rounds per minute) and low (4,000 rounds per minute). A round is fed, chambered, fired, extracted, and ejected from each of the six barrels on each revolution.

▲ The last A-7E squadrons in Navy service, VA-46 and VA-72, provided support for Desert Storm from USS *John F. Kennedy*. Desert Storm operations generated 725 sorties averaging 4.3 hours each, for a total of 3,100 combat flight hours, during which A-7s carried a variety of ordnance and also served as tankers. No A-7s were lost to enemy action during the entire operation. The two squadrons returned to Cecil Field, Florida, where they were decommissioned on 23 May 1991. (US Navy)

◀ A-7Es of VA-46 refuel from an A-6 configured for the tanker mission. They are armed with Rockeye CBUs, a primary anti-personnel munition used to attack Iraqi Republican Guards positions in Kuwait. (US Navy)

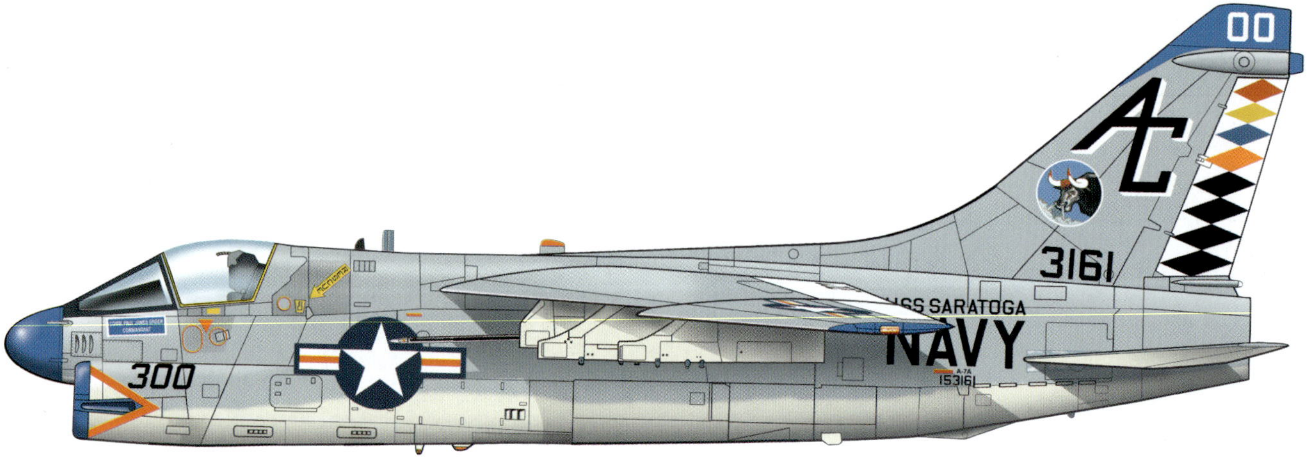

◂ A-7A, BuNo 153161, CVW-3 CAG aircraft of VA-37 'Bulls,' USS *Saratoga* (CVA 60), 1970. The Corsair II is painted in the basic carrier aircraft color scheme used by the US Navy from mid-1951 to the late 1980s: gloss Light Gull Gray (FS595a: 16440) upper surfaces and gloss White (FS595a: 17875) lower surfaces and control surfaces (ailerons, flaps, stabilators, and rudder).

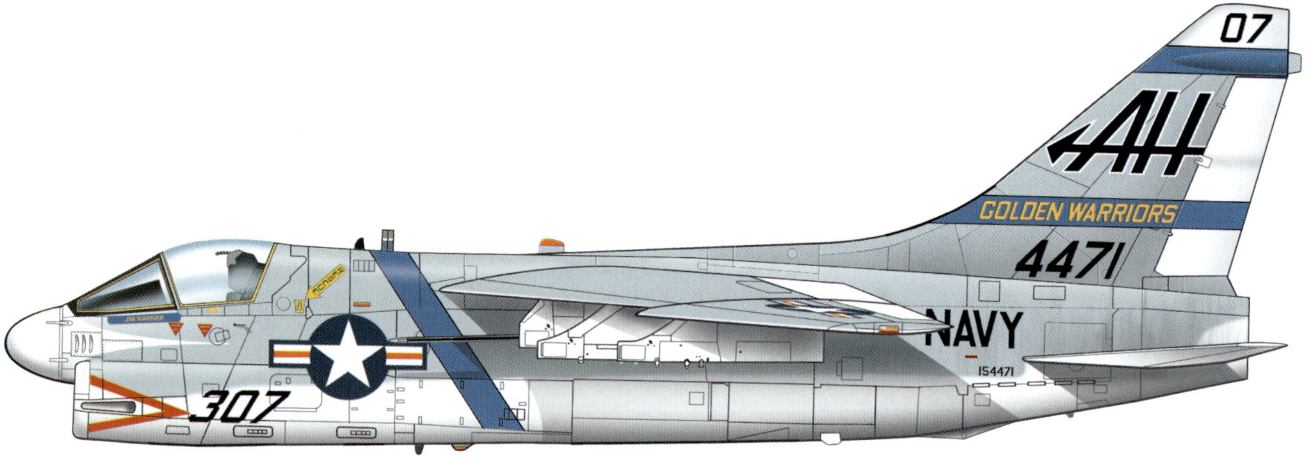

◂ A-7B, BuNo 154471, of VA-87 'Golden Warriors,' c. 1970, in the basic light Gull Gray and White carrier aircraft color scheme. Tail stripes are light blue instead of orange as usually seen on Golden Warriors' aircraft. This A-7B was one of twenty-four converted to two-seat TA-7Cs.

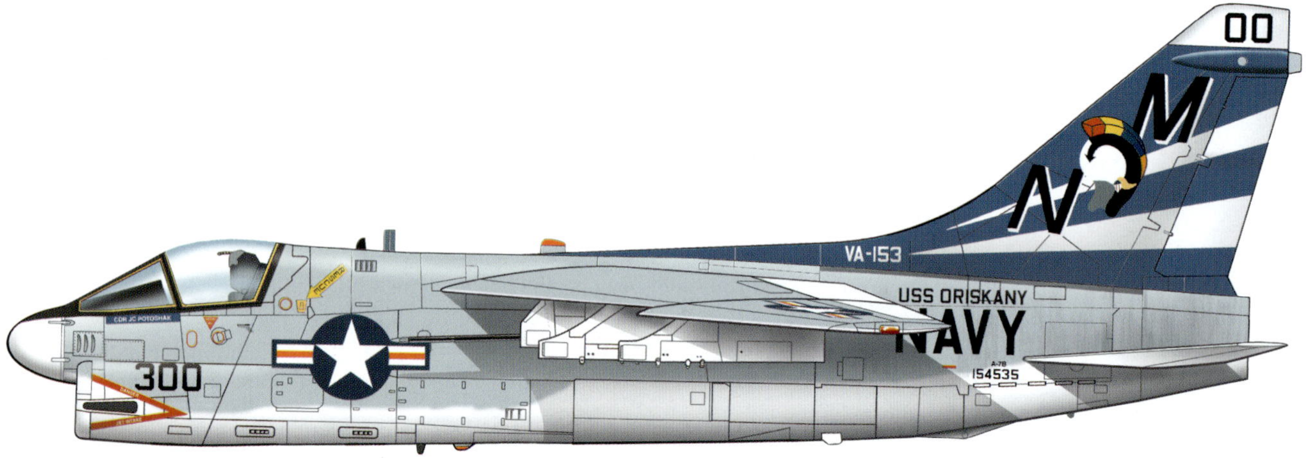

◂ A-7B, BuNo 154535, of VA-153 'Blue Tail Flies,' USS *Oriskany* (CVA 34), 1970, in the Light Gull Gray and White color scheme.

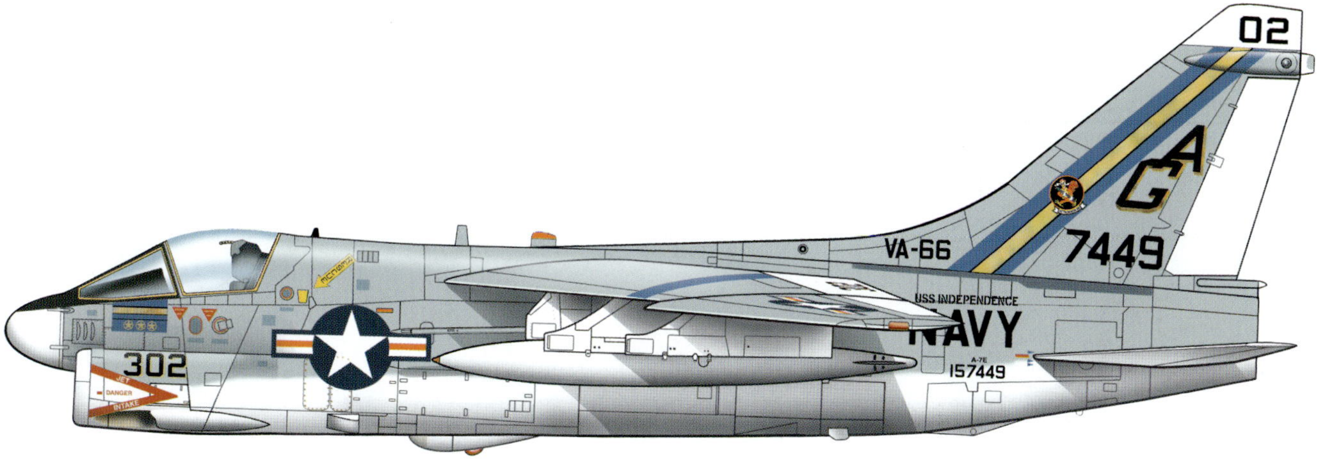

A-7E, BuNo 157449, of VA-66 'Waldos,' USS *Independence* (CVA 62), 1975, in the basic carrier aircraft color scheme.

A-7E, BuNo 158013, of VA-22 'Fighting Redcocks,' USS *Enterprise* (CVN-65), 1986. This Corsair II was painted in an experimental 'wrap-around' color scheme of Black (FS595a: 37038), Tan (FS595a: 30219), and Gray (FS595a: 36622). Markings are Black and Light Gull Gray (FS595a: 36440). The Black very quickly faded to a charcoal gray shade.

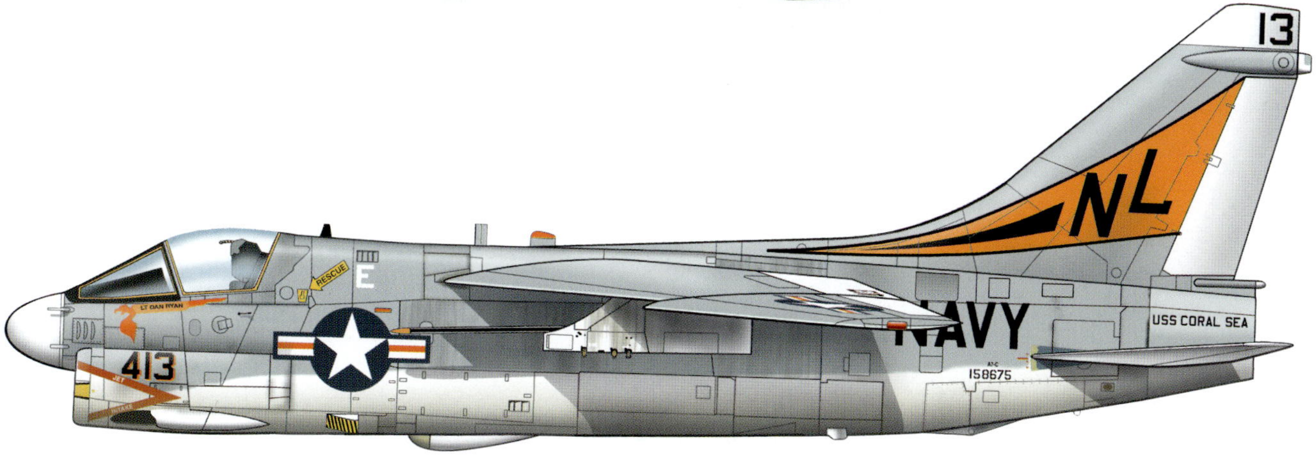

A-7E, BuNo 158675, VA-94 'Mighty Shrikes,' USS *Coral Sea* (CVA 43), 1971. Basic color scheme is Light Gull Gray and White. VA-94 was the first squadron to deploy with the A-7E.

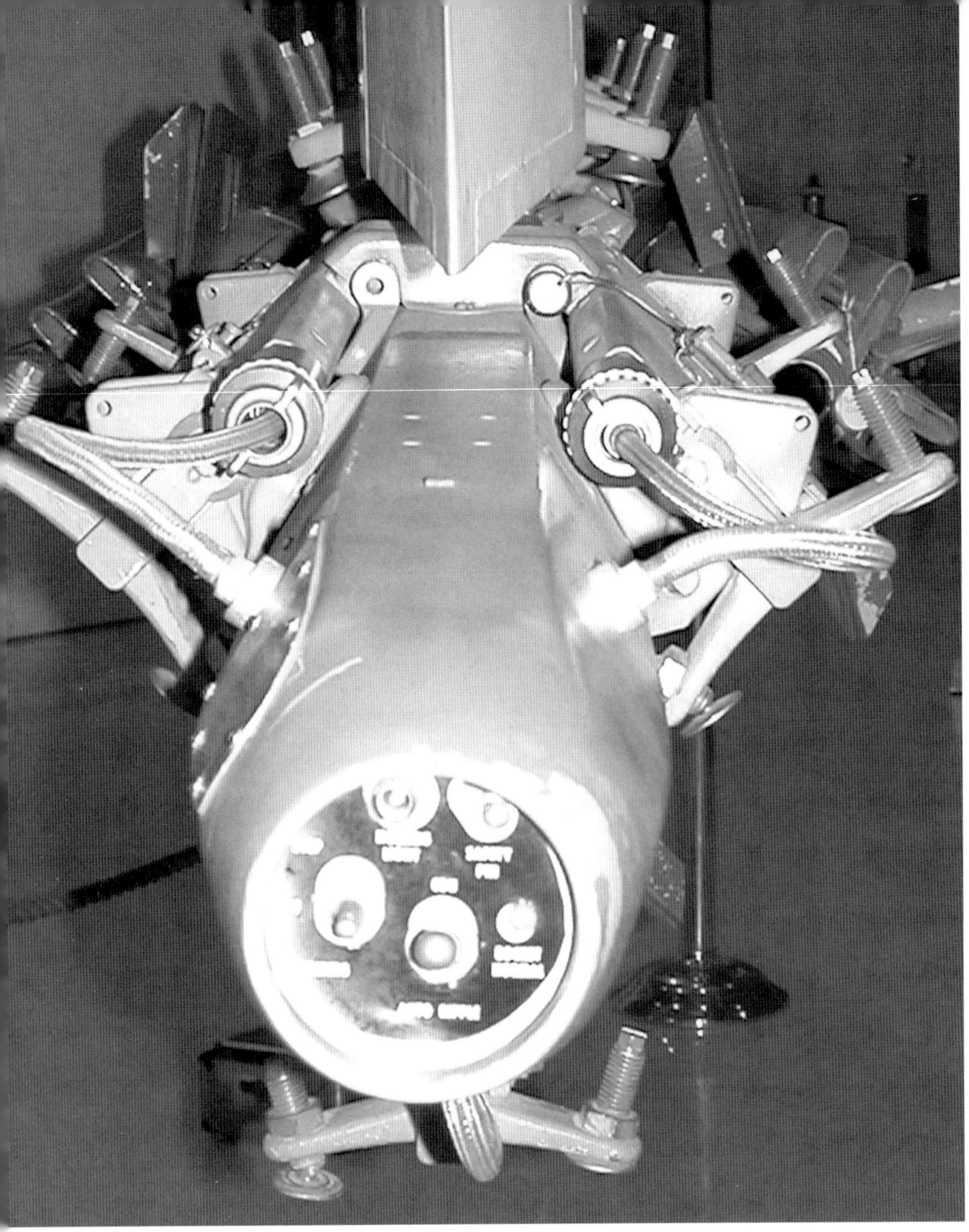

▲ Aft end of the Multiple Ejector Rack (MER), mounted on an A-7E. (Lou Drendel)

▲ An A-7E on a Desert Storm mission refuels from a USAF KC-135 configured for tanking Navy aircraft. The Navy uses the probe-and-drogue system, requiring tanker aircraft to be equipped with the basket-shaped drogue. This Corsair II is armed with AGM-84 High Speed Anti-Radiation (HARM) missiles, used to attack enemy surface-to-air missile (SAM) sites. (USAF)

Ground Clearance

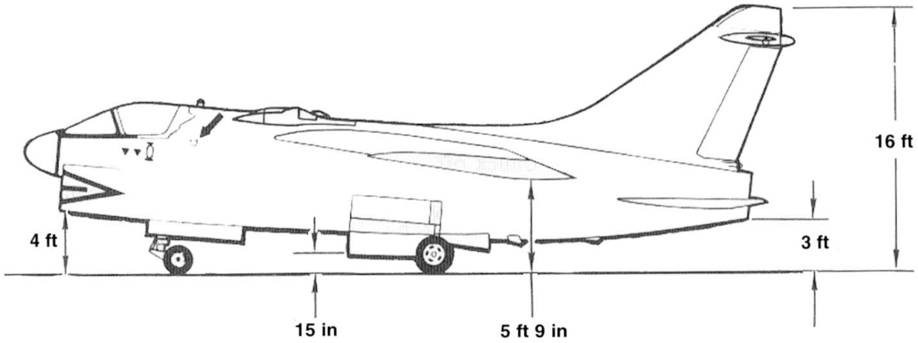

◄ MER with Mk 82 Snakeye Low Drag General Purpose (LDGP) bombs mounted. The Mk 80 series was developed in the 1950s in response to the need for bombs with less aerodynamic drag. All Mk 80 series bombs have similar construction, are cylindrical in shape, and are equipped with conical fins or retarders for external high-speed carriage. They are fitted with both nose and tail fuses to ensure reliability and produce effects of blast, cratering, or fragmentation. During Desert Storm, the Mk 80 series of bombs were dropped from literally every fixed-wing aircraft that supported the ground offensive. (Lou Drendel)

Mixed Store Configurations

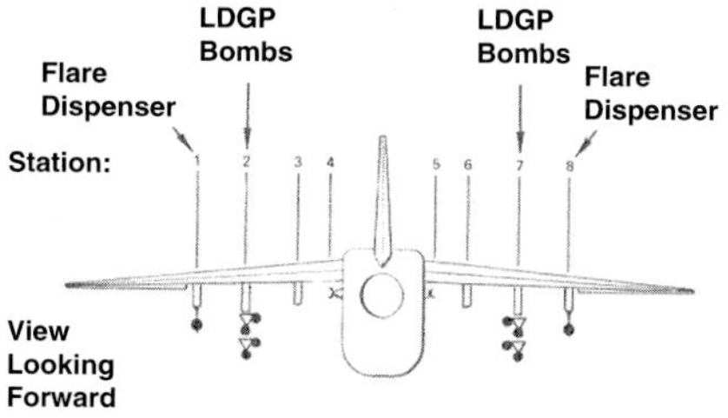

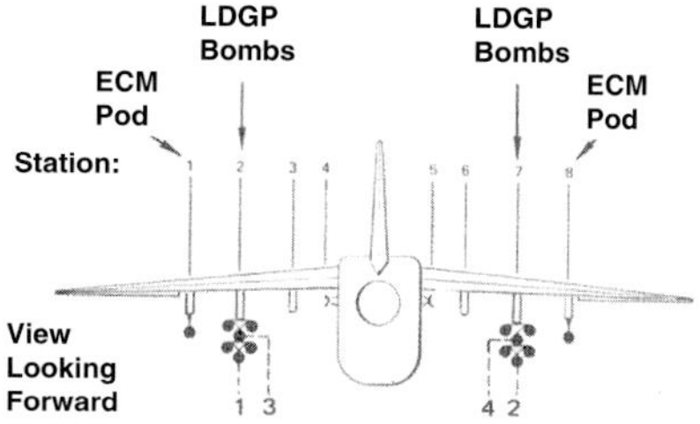

NORMAL RELEASE SEQUENCE

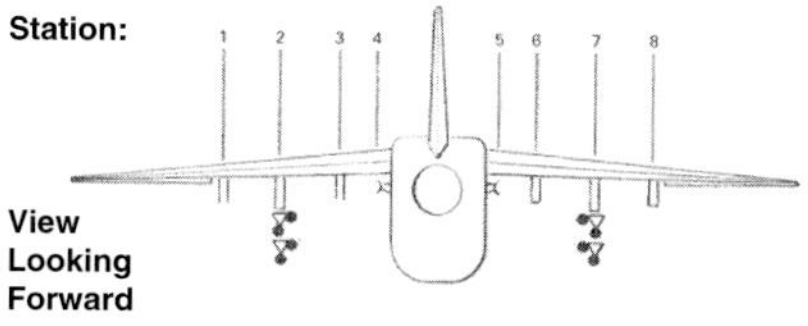

CASE 1 – ALL FLARE DISPENSERS DELETED TO OBTAIN AN ADDITIONAL LDGP BOMB LIKE STORE CONFIGURATION

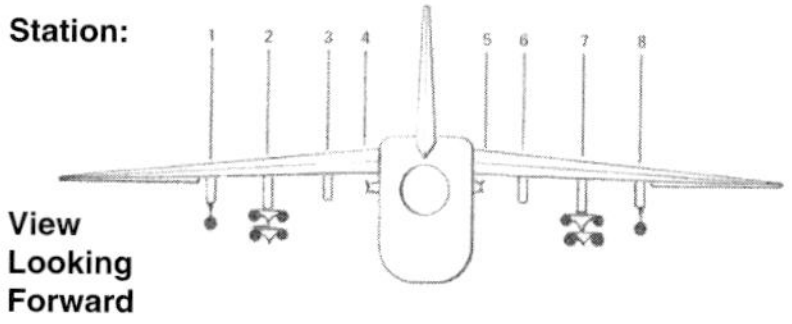

CASE 2 – FOUR LDGP BOMBS DELETED IN NORMAL RELEASE SEQUENCE TO OBTAIN A PARTIAL MIXED STORE CONFIGURATION

▼ A-7D of the 152nd Tactical Fighter Training Squadron (TFTS) of the 162nd Tactical Fighter Training Group (TFTG) of the Arizona Air National Guard at Tucson, Arizona. (Andre Jans)

▲ The A-7Ps served for a time with the FAP in the interceptor role and carried AIM-7P Sidewinder air-to-air missiles. Before their retirement, they provided tactical air support for maritime operations and offensive and defensive air support with a range of iron bombs. They also were equipped to carry the AGM-65A Maverick air-to-surface missile. They were provided with rear warning receivers and chaff/flare dispensers, and could carry AN/ALQ-131 jamming pods. The A-7Ps suffered a relatively high attrition rate in FAP service, and spares support was provided by twenty non-flyable ex-USN A-7As. (Jorge Manuel Antao Ruivo)

◄ Portuguese A-7P serial A-043 (former US Navy A-7A BuNo 153134) was delivered in 1984 and went out of service in 2000. (Jorge Manuel Antao Ruivo)

▼ The first USAF A-7Ds were deployed from Myrtle Beach AFB, South Carolina, to Korat Royal Thai Air Base in mid-October 1972. This A-7D carries a TV-guided Maverick missile. After the protective cover is automatically removed from the missile's nose and its video circuitry is activated, the scene viewed by the guidance system appears on a cockpit television screen. The pilot selects the target, centers cross hairs on it, locks on, then launches the missile.

▲ The 'Pave Penny' system, or Target Indicator System Laser (TISL), was added to 383 A-7Ds, installed within a chin fairing under the intake. It is slaved to the HUD, improving target acquisition and bombing accuracy.

▶ Captain Don Cornell in the 'last chance' arming area at Korat RTAB. The wing commander's airplane was the only A-7D to carry the sharkmouth, which matched the markings on the SAM-hunting F-105Gs of the 17th Wild Weasel Squadron. (Chuck deVlaming via Don Cornell)

▼ During the two-and-a-half months that the 354th was in combat in 1972, the wing's seventy-two aircraft averaged sixty-two missions per day, producing a total of 6,568 sorties in 16,819 combat flying hours. Their radius of action averaged 350 miles and extended as far out as 550 miles.

The 'Royal Maces' of VA-27 were commissioned on 1 September 1967 flying the A-7A, reporting to Carrier Air Wing 14 in January 1968. They made their first combat deployment aboard USS *Constellation* (CVA 64) in May 1968. During a second combat cruise aboard *Constellation*, beginning in August 1969, VA-27 flew over 2,500 combat sorties. This Corsair II is armed with AIM-9B Sidewinders on the fuselage stations. (US Navy)

Weapon Release Controls

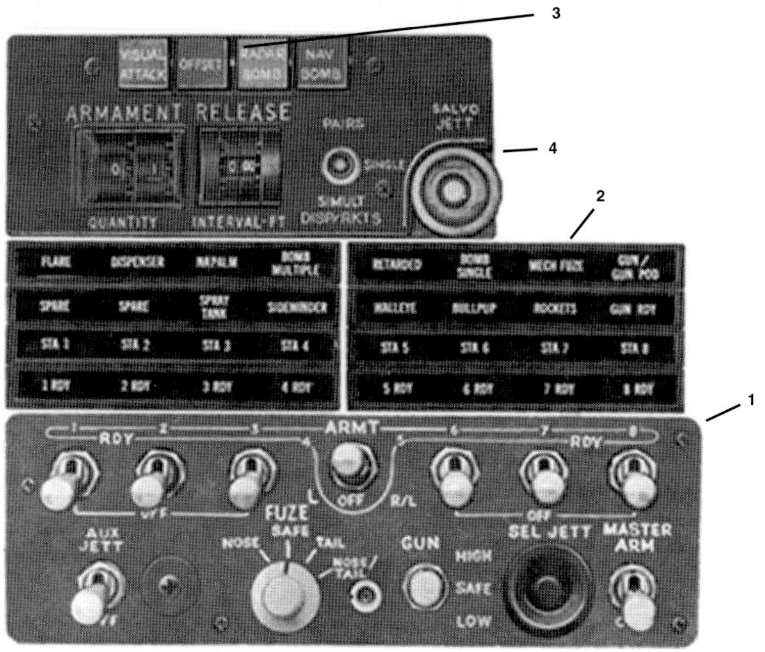

1. Armament select panel
2. Armament advisory light panel
3. Attack switches
4. Armament release panel
5. Armament release switch
6. Trigger switch

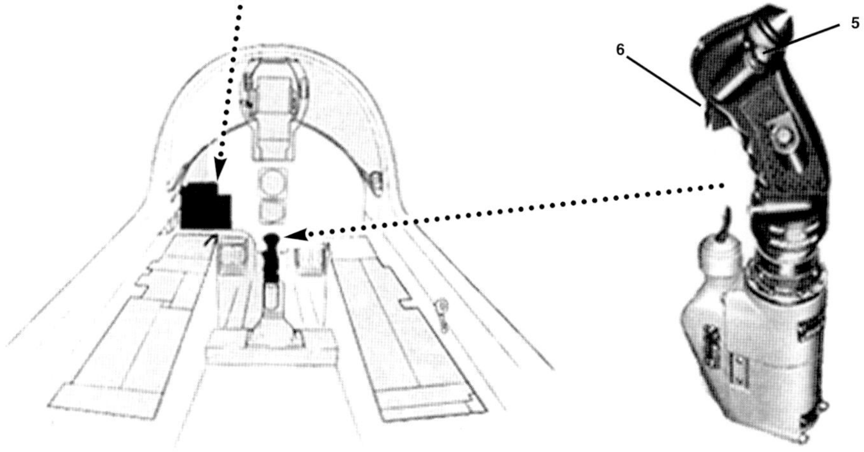

▲ A-7E, aboard USS *John F. Kennedy*, loaded with five Rockeye cluster bomb units (CBUs). The Mk 20 Rockeye is a free-fall, unguided cluster weapon designed to kill tanks and armored vehicles. The system consists of a clamshell dispenser, a mechanical Mk 339 timed fuse, and 247 dual-purpose armor-piercing shaped-charge bomblets. The bomblet weighs 1.32 pounds and has a 0.4-pound shaped-charge warhead of high explosives, which produces up to 250,000 psi at the point of impact, allowing penetration of approximately 7.5 inches of armor. (US Navy)

▼ A-7 pilot confers with plane captain prior to a Vietnam Alpha Strike. (US Navy)

A-7E of VA-46 loaded with Mk 80 series bombs for a Desert Storm mission. Some bombs are thermally protected for use on aircraft carriers. The thermally protected Mk 80 series bomb was developed to increase the cook-off time and decrease the reaction of bombs when engulfed in a fuel fire. The Mk 82 and Mk 83 series LDGP bombs underwent a Product Improvement Initiative (PII) which entailed filling the bomb cases with a less sensitive explosive. When so filled, the Mk 82 and Mk 83 bombs are redesignated BLU-111/B and BLU-110/B, respectively. They are recognized by their rougher exterior casings. (US Navy)

▲ The final operational version of the Corsair II was the two-seat A-7K, which never served with the active duty USAF, but went directly to the Air National Guard. The first A-7K (73-1008), which made its maiden flight in January 1981, was a conversion of an existing A-7D, while a further thirty airframes were built as two-seaters. (Andre Jans)

◄ Triple Ejector Racks (TERs) ready to be fitted to Navy attack aircraft. (Andre Jans)

▼ A-7D of the 146th Tactical Fighter Squadron, 112th Tactical Fighter Group, Pennsylvania Air National Guard, based at Pittsburgh. The blade antenna which appears on the fairing above the rudder on A-7D/K aircraft is for the VOR/TACAN system. (Peter Steendam)

◂ A pair of A-7Ds of the 354th TFW on approach for a formation landing. Note that the wingman is stacked slightly below the leader, and if he maintains this position, he will touch down before the lead aircraft. Both aircraft carry 300-gallon external tanks and MERs on wing pylons. (Shinichi Ohtaki)

▾ A-7A of VA-147 about to launch, configured for the Wild Weasel SAM killing mission. It is armed with CBUs for killing the SAM site operators, and an AGM-45 Shrike missile. The AGM-45 was the first missile built specifically for the anti-radar mission, and more than twenty thousand were produced beginning in 1962. The Shrike's effectiveness was limited by the requirement that it be pointed at its intended target radar during launch, and the fact that the Shrike would lose its lock if the radar ceased to radiate. (US Navy)

▾ An A-7E 'bolters' on USS *America*, 1985. A 'bolter' occurs when the arresting hook skips over the wire and fails to engage. This is why carrier aircraft go to full power as soon as they hit the deck, making a touch-and-go possible. (Lou Drendel)

▲ An A-7E of VA-81 early in the application of low visibility tactical camouflage. It still carries the full size national insignia and gloss white 300-gallon external tank, though all other markings on this CAG bird are subdued.

▼ An A-7A of VA-147 gets the two-finger run-up signal from the Catapult Officer aboard USS *Ranger* (CVA 61) during the first combat cruise of the Corsair II. It is loaded with Mk 80 series Snakeye high drag bombs. The folding fin Snakeye has been replaced by a ballute system retarder on high drag bombs. (US Navy)

'Sharkmouth' A-7D, serial 70-0982, 3rd TFS, operated from Korat Royal Thai Air Base, Thailand, from 1973-1975. Camouflage is the standard USAF Southeast Asia or 'Vietnam' scheme: upper surfaces in Tan (FS595a: 30219), Green (FS595a: 34079), and Dark Green (FS595a: 34102), with under surfaces in Gray (FS595a: 36622). This Corsair II is currently on display at Volk Field, Wisconsin, in the markings of the Puerto Rico Air National Guard.

A-7E BuNo 160552 of VA-72, USS *John F. Kennedy* (CV 67), in Operation Desert Storm markings. During Desert Storm, the squadron's Corsair IIs participated in both the first combat strike against Baghdad and the last naval air combat strike of the war, hitting retreating Iraqi troops east of An Najef, Iraq. During the forty-three days of war, VA-72 flew 362 sorties without the loss of a pilot or aircraft. The Corsair II is depicted here in flat Ghost Gray (FS595a: 36375) overall with black and white markings and blue and white rudder checks. It was later painted in a two-tone scheme of flat Tan (FS595a: 31667) and flat Brown (FS595a: 30233).

A-7E BuNo 156889 of VA-195 'Dam Busters,' in the basic light Gull Gray and White carrier aircraft color scheme, but marked as the CAG's airplane. During the Vietnam War, as a unit of Carrier Air Wing 11 on USS *Kitty Hawk* (CV 63), VA-195 delivered the first data link version of the television-guided Walleye glide bomb in combat. On July 19, 1972, the squadron delivered a single Walleye down the throat of a cave storage area, causing its complete destruction. Later the same day, they destroyed the Ninh Binh railroad bridge, also with a single weapon.

The 'Ravens' of VA-93 carried this very impressive sharkmouth on their A-7s into combat in Southeast Asia, flying off USS *Midway* as part of CVW-5 from 1971 to 1973. Basic color scheme is Light Gull Gray over White.

CAG A-7E, BuNo 157480, of VA-83 landing with stabilator in the full nose-up position. The 'Rampagers' flew A-7Es from 1971 to 1987 as part of CVW-17, operating primarily in the Mediterranean off carriers USS *Forrestal* (CV 59) and USS *Saratoga* (CV 60). Basic color scheme is Light Gull Gray over White.

A-7E BuNo 159968, CAG aircraft of VA-87, comes to a full stop in the arresting gear of USS *Independence* (CV 62) somewhere in the Mediterranean. Following its establishment in 1968, the 'Golden Warriors' flew both A-7Bs and A-7Es until 1985, when they traded their Corsair IIs for F/A-18 Hornets. Basic color scheme is Light Gull Gray over White.

▲ An A-7B of VA-105 armed with 2.75-inch rocket launchers on Station 8.

▶ A-7Bs of VA-82 await loading with M117 750-pound bombs. The basic M117 dates from the Korean War and uses a low-drag tail fin for medium and high-altitude deliveries.

▼ An A-7E of VA-27 armed for a Linebacker II anti-SAM mission with an AGM-45A Shrike missile. The AERO-5B-1 and LAU-118 series rail launchers are used to launch the Shrike guided missile and provide the electrical and mechanical interface between the missile and the launch aircraft. VA-27 was one of the longest-serving Corsair II squadrons. It was established on 1 September 1967 and flew A-7s until being redesignated Strike Fighter Squadron (VFA) 27 on 24 January 1991, equipped with F/A-18 Hornets.

▲ The Ohio Air National Guard operated two A-7D/K squadrons. The 112th TFS of the 180th TFG at Toledo, and the 162nd TFS of the 178th TFG at Springfield. In 1977 the USAF began retrofitting its A-7D fleet with Automated Maneuver Flaps (AMF) to improve the aircraft's performance at high angles of attack and to counter its tendency to depart from controlled flight at excessive angles of attack. (Andre Jans)

▼ A-7D of the 355th TFW at Davis-Monthan AFB, Arizona. Unique to the USAF's A-7D were the boom-and-receptacle in-flight refueling system; higher energy-rated wheels, tires, and brakes; and the ESCAPAC IC ejection seat, modified to utilize the USAF survival kit and restraint system together with a low-pressure, demand-type oxygen system. (Don Logan)

▲ Mk 82 500-pound bombs being loaded on an A-7A for a Vietnam combat mission. They are equipped with the Mk 15 Snakeye folding fin retarder system. The A-7 is also loaded with early model AIM-9 Sidewinder missiles. The AIM-9 has been in service since 1956, but current models are several magnitudes greater in operational effectiveness than the AIM-9B on this Corsair II. (US Navy)

◀ 'Mighty Mouse' nose art on an Air National Guard A-7D. Hallmark of the A-7D/E series was a state-of-the-art navigation/weapon delivery system built around the ASN-91 central computer. Major systems elements included the APQ-126 radar, AVQ-7 head-up display, and a projected map display system. There were also major improvements in the Doppler radar, inertial navigation system, and weapon control system. (Andre Jans)

▼ The LAU-7/A launcher provides a complete launching system for the AIM-9 Sidewinder. LAU-7A (series) is a reusable single rail launcher that provides the mechanical and electrical interface between the missile and the launch aircraft. It houses the nitrogen receiver assembly used to cool the missile guidance system's infrared detector. (Dave Mason)

▲ Portuguese A-7Ps were finished in a wrap-around version of the USAF's Vietnam camouflage scheme. (Peter Steendam)

▲ The principal feature of the YA-7F was incorporation of a Pratt & Whitney F100-PW-220 afterburning engine of 20,000 pounds thrust. This required modification of the fuselage to add 29½ inches just forward of the wing, 18 inches immediately aft of the wing, and upward rotation of the aft fuselage to maintain ground clearance.

▼ A-7D of the 149th TFS, 192nd TFG, Virginia Air National Guard, which flew the A-7D from 1982 to 1992. The two-tone blue camouflage was the final USAF markings scheme of the Corsair II's career. (Peter Steendam)

◂ A-7D of the 3rd Tactical Fighter Squadron in the last-chance arming area at Korat RTAB. The 3rd TFS assumed the Corsair II combat role after the 354th Wing returned to Myrtle Beach in 1973. The 3rd remained on site and flew missions in support of Operation Frequent Wind (the evacuation of Saigon) and the Mayaguez rescue missions in 1975. This A-7D is loaded with Mk 83 LDGP bombs. (USAF)

▾ A-7D of the 354th TFW in a revetment at Korat RTAB during the Linebacker campaigns of 1972. During the Vietnam War, 12,928 A-7D combat sorties were flown, with only four losses. The last US air strike into Cambodia was made by an A-7D on August 15, 1973.

▾ Ordnance technician attaches a fuse to a Mk 82 bomb on a MER. General purpose bombs may use both nose and tail fuses and conical or retarded tail fins. The bomb is usually equipped with the mechanical M904 (nose) and M905 (tail) fuses. (US Navy)

▲ A-7Ds of the 3rd TFS undergo engine maintenance at Korat RTAB in 1973. The Allison TF41-A-1 is a twin spool, axial flow, turbofan engine. The non-afterburning TF-41 was a license-built version of the Rolls Royce Spey engine producing 14,500 pounds of thrust. (USAF)

◀ An A-7D of the 174th TFS, 185th TFG, Iowa ANG, based at Sioux City, Iowa. It carries a pair of Mk 84 2,000-pound LDGPs and an ALQ-119 ECM pod. The Pave Penny pod (AAS-35) under the nose is a receiver which obtains information from a designator then displays it on the HUD. (Ted Carlson)

A-7H 159913 of the Hellenic (Greek) Air Force taxis out for takeoff. The two squadrons of the 116th Combat Wing fly A-7Es from Araxos Air Base, and another two squadrons of the 115th Combat Wing are equipped with the A-7H variant at Souda air base on Crete. Greek A-7s carry US Navy serial numbers, commonly referred to as 'BuNos' (Bureau of Aeronautics numbers). (Colin Norwood)